P9-CBV-715

INTERSTATE

**The Roads that Built America**

# The Roads that

**DAN McNICHOL**

# Built America

## The Incredible Story of the
## U.S. Interstate System

MAIN LIBRARY
CHAMPAIGN PUBLIC LIBRARY
505 SOUTH RANDOLPH STREET
CHAMPAIGN, ILLINOIS 61820-5193
09/08/2006

To Jin
—Dan McNichol

# The Roads that Built America

Published by Sterling Publishing Co., Inc.
387 Park Avenue South, New York, NY 10016
© 2006 by Sterling Publishing Co., Inc.
Distributed in Canada by Sterling Publishing
c/o Canadian Manda Group, 165 Dufferin Street
Toronto, Ontario, Canada M6K 3H6
Distributed in Great Britain by Chrysalis Books Group PLC
The Chrysalis Building, Bramley Road, London W10 6SP, England
Distributed in Australia by Capricorn Link (Australia) Pty. Ltd.
P. O. Box 704, Windsor, NSW 2756, Australia

10  9  8  7  6  5  4  3  2  1

Manufactured in China.
All rights reserved

Photography and illustration credits are found on page 271
and constitute an extension of this copyright page.

Sterling ISBN 1-4027-3468-9

For information about custom editions, special sales, premium and
corporate purchases, please contact Sterling Special Sales
Department at 800-805-5489 or specialsales@sterlingpub.com.

Design:  Richard J. Berenson
         Berenson Design & Books, LLC, New York, NY

# Contents

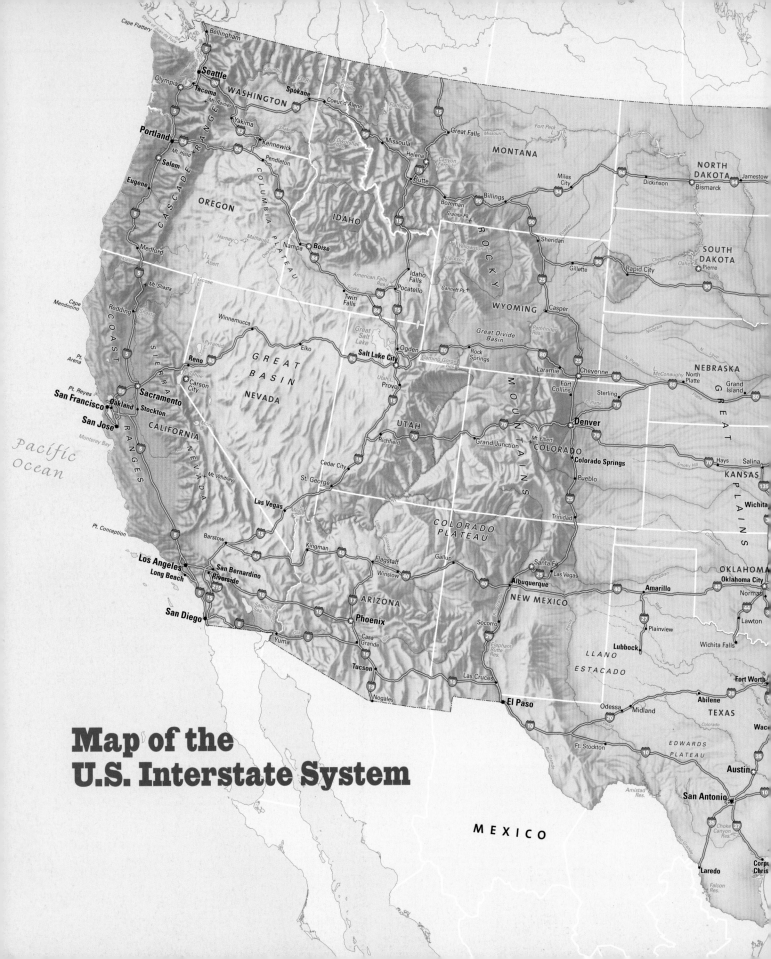

# Map of the
# U.S. Interstate System

The U.S. is a nation connected like no other. The Interstate System binds 48 states and their capitals to one another. Shown here is Houston's I-610.

CHAPTER ONE

# The Roads that Built America

**W**HETHER PEOPLE YOU LOVE are about to take a road trip across the country or just run an errand across town, tell them to stay off the local roads and on the Interstate System. They'll get where they're going faster, and they're twice as likely to make it there alive. It's a fact.

The Interstate System, officially known as the Dwight D. Eisenhower System of Interstate and Defense Highways, is capable of taking us to every corner of the United States. Its highways are the safest in the country by a factor of two, and they carry us where we are going faster than any other road in the nation. The U.S. Interstate System is unique. There is nothing like it in the world.

We are the only country on earth with such a vast network of superhighways. Not Russia nor China nor all the European nations combined can match it. It is the largest single engineering and construction project on this planet. In scale it is far larger than the Great Pyramids of Egypt, the Great Wall of China, the aqueducts of Rome, and the Suez and Panama canals.

But America is often dismissive of her grandest works. When greatness comes often, it is easy to be indifferent about your accomplishments. The Interstate System is a prime example of greatness taken for granted, of overachievement as the norm.

**In 1990,** the Interstate System was renamed to honor its principal visionary.

There are 62 routes on the Interstate System. Of these, only three are transcontinental highways running from coast to coast (I-10, I-80, I-90). However, seven Interstates run between Canada and Mexico (I-5, I-15, I-35, I-55, I-65, I-75, I-95).

Here at home it has been ridiculed as often as it has been celebrated, while the rest of the world marvels at it. The Interstate System is a phenomenon that deserves a long look and a lot of praise.

The roots of the System are militaristic and defensive, with a large dose of national pride thrown in. It is the result of the leadership of our 34th President, Dwight D. Eisenhower, although many other Presidents saw the need for a national system of roads, including George Washington.

But it was Eisenhower who envisioned the broad "ribbons across the land," who made the Interstate System his favorite domestic agenda and persuaded Congress to supply the money for it. Eisenhower had witnessed firsthand the power and importance of good roads. As Supreme Commander of the Allied Forces during World War II, he had used the Germans' own superhighway, the Autobahn, to track down and defeat their army. As President during the Cold War of the '50s, he was consumed with the threat of nuclear attack. He saw an Interstate System of highways as essential to the nation's defense, needed for the fast deployment of troops and as a possible escape route for Americans fleeing the fallout of an atomic blast.

The building of the System began in 1956, shortly before Ike's second term. The first ground broken was in Missouri, beginning construction on what is now I-70. Over the next half century, the Interstate System has been a work in progress. The Big Dig in Boston is the last chapter in the construction phase of the system. When that gigantic piece of work is finished, I-90 will at last complete its intended journey from coast to coast and the building of the Interstate System will be over. All subsequent work will be considered repair or reconstruction. Or at least that's the plan at the moment.

## WHAT EXACTLY IS THE INTERSTATE SYSTEM?

The Interstate System is 42,795 miles of the nation's primary highway network. Imagine looking at a road atlas of the continental United States, with only the Interstate routes showing. You would see 62 superhighways crisscrossing the nation in a grid. Twenty-seven of them are labeled with even numbers, and the traffic on them travels east and west. The remaining 35 routes are marked with odd numbers, and traffic on them flows north and south. In addition to these major routes, there are 261 beltways and spur roads circling or beelining into and out of all the major urban areas across the country.

The genius of the Interstate System is in its uniformity. Wherever you find yourself along it, there are at least two travel lanes in each direction, and each lane is 12 feet wide. On the far left side of the road, drivers have a four-foot-wide shoulder, and to the far right they have a 10-foot-wide breakdown lane.

In the hinterlands, banked curves are designed to keep a car safely on the road at 70 miles per hour. In the city, curves are engineered to handle traffic moving at 50 miles per hour. Curves rarely exceed a one percent curvature. The steepness of the Interstate is not to exceed a six percent grade in mountainous regions or five percent in town. The System's highways are straighter, flatter, and faster than almost any other road.

Everything on the Interstate is intended to protect your life. Guardrails on hazardous stretches, berms or concrete dividers in the median, landscaping that screens oncoming lights, breakaway signs and lampposts all save lives. Even uniformity of signage prevents sudden and potentially deadly pulls of the wheel from a startled or confused driver.

Without an Interstate System, 6000 more people would lose their lives in automobile accidents every year. Nearly two people die for every 100 million miles traveled on other highways, compared to less than one person for every 100 million miles traveled on the Interstate.

**Rumble strips:** The vibration created when a vehicle drifts off the road and over a Rumble strip saves lives by shaking sleepy drivers. An inventor whose son was killed in such an accident helped implement their usage.

## RIDDEN HARD

Since 1956, the year ground was broken on the Interstate System, America has experienced a transportation boom. Call it the freedom of mobility. We are moving about the nation as never before, living and working in areas that previous generations considered too distant to be practical. Americans put 340 percent more miles on their vehicles each year than they did 50 years ago. At any given moment, 50 percent of all the large commercial trucks in the United States are hauling their cargo on the Interstate. The System's designers have planned for this boom. Each superhighway is designed to handle traffic volumes 20 years out from its date of construction or improvement. Each of the 54,663 bridges and 104 tunnels is designed to handle traffic for 40 years from its creation.

The highways have fulfilled their original purpose: to reduce travel times, improve commerce, and protect our nation from military aggression. And they have done more. The System is our major, sometimes the only, link to all the other worlds of transportation. Airports, seaports, train stations, other highways, byways, and local streets interlock with the Interstate System. It is crucial to our nation's economic strength and social harmony.

## UNITED WE STAND

But Ike was the consummate military man. The Interstate System he pushed through Congress is also the center of the nation's Strategic Highway Network (STRAHNET), the core of our domestic land-based defense network. Every one

*On September 11, when the twin towers of the World Trade Center in lower Manhattan fell to terrorism, the Interstate was needed. And it performed brilliantly.*

of the country's 200 military bases has an efficient connection to the Interstate. Its superhighways can accommodate emergency landings and takeoffs of airplanes and the movement of vast numbers of men and matériel. Much of our military equipment is specifically designed to be moved along the Interstate. Every city of any size has emergency evacuation procedures built around the System. Neither the cities nor the federal government talk much about the protocols that are in place, but they are there.

Fortunately for our nation, we have not had to call upon these procedures for any full-scale national emergency. But on September 11, 2001, when the twin towers of the World Trade Center in lower Manhattan fell to terrorism, the Interstate was needed. And it performed brilliantly.

The collapse of the towers created a 1,000,000-ton pile of steel, glass, and concrete. Unknown thousands of people were missing. Rescue workers were desperately seeking a piece of equipment that could cut through high piles of jagged steel I-beams and concrete-reinforcing bars to search for survivors.

Within 24 hours their efforts led them to Aurora, Illinois, and a 345 Ultra High demolition machine developed by Caterpillar. With powerful hydraulic shears and grapplers and an 85-foot-high reach, it was without question the machine the rescue workers wanted. But getting it to Lower Manhattan from Illinois was the challenge.

Caterpillar crews began scrambling. First order of business: break down the huge machine into shippable parts. Meanwhile other personnel were on the phone with the Illinois State Police. They were performing a homespun version of what the Department of Defense must do when hauling matériel to seaports during mobilization—checking with all the states to find out the latest height and weight restrictions under and over the Interstate System.

The difference was that this precious cargo was going twice as far as the army normally ships. Escalating matters was the fact that special exceptions were needed. Because the 345's arrival was a matter of life and death, it would have to move day and night over the Interstates at a weight, height, and width that were considered oversize by every state it had to pass through. Nonetheless, all the states—Illinois, Indiana, Ohio, Pennsylvania, New Jersey, and New York—immediately cut through their red tape and made the movement possible.

Back in Illinois, working all night, crews dismantled the 345 into three parts and loaded it onto three trailers, each one longer than two Greyhound buses end to end. Early in the morning of September 13, accompanied by two escort vehicles with flags waving and lights flashing, the three big tractor-trailers

To build the Interstate System took enormous quantities of cash. Billions of dollars from user taxes on fuel, tires and other highway-related products were funneled into the Highway Trust Fund for construction. The Fund has remained safe from would-be "Breakers of the Trust" who have periodically tried to raid it for transit projects and the government's general budget.

**In the hours after** the terrorist attacks of September 11, 2001, the 345 Ultra High and other rescue equipment were rushed along the Interstate System to New York City and Washington, D.C., in the hope of saving lives.

headed out. They made their way to I-80 and then, driving through the night, the convoy passed over five different Interstate highways. Within 20 hours they were in New York City, where search-and-rescue teams were anxiously waiting.

When the 345 arrived, it was stationed at the highest corner of the wreckage. With its long reach and powerful jaws, it began its sad task of gently sifting through the rubble searching for survivors and, later, simply hoping to recover remains. Caterpillar shipped the 345 and the men who could operate it with no discussion of payment. It was their donation to the recovery operation.

The terrorist attack of 9/11 was a strike against U.S. commerce as well as a strike against our way of life. It was meant to paralyze trade and break the nation's spirit. But that never happened. For weeks after the attacks, special salvage equipment was rushed along the Interstate System to New York City and Washington, D.C., from as far away as Georgia and Michigan. Around the nation, planes grounded in the wake of the attack discharged their payloads to waiting trucks, which sped critical shipments across the most amazing network of superhighways the world has ever known—the Dwight D. Eisenhower System of Interstate and Defense Highways.

Eisenhower was a military man, but he was also a visionary. He believed the country needed a network of superhighways not only for military purposes but to create a lasting bond across the nation. "Together, the uniting forces of our communication and transportation systems are dynamic elements of the very name we bear—United States," he wrote. "Without them, we would be a mere alliance of many separate parts."

Ike would be deservedly proud of the Interstate System that bears his name.

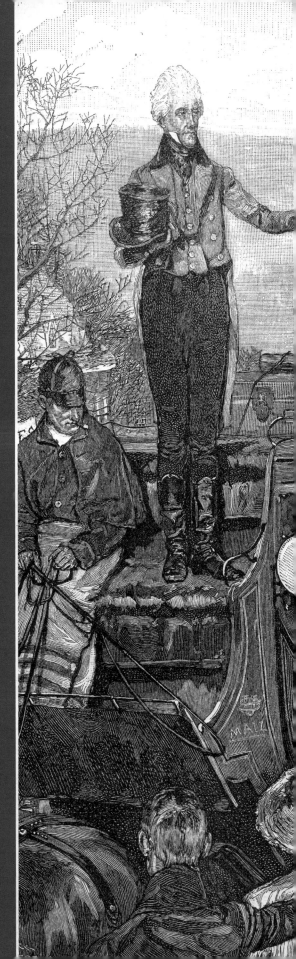

President-elect Andrew Jackson, in 1829, gathered the populace as his stagecoach made its way along the National Road, America's first interstate.

CHAPTER TWO

# America's First Interstate

**I**N 1784, JUST A YEAR after the end of the Revolutionary War, President George Washington set out on horseback from his Mount Vernon home, riding at his normal pace of five miles per hour. This was no casual early morning canter. Washington was on a mission. He wanted to build a road, a national road, to keep his young country united. Traveling into the Ohio country and covering 680 miles in five weeks, he surveyed the land, and its people, for the best route.

The picture on the Western Frontier was grim. American settlers were trapped between two world powers. Washington feared they would fall into the control of "the Spaniards on their right or Great Britain on their left." He declared that the nation should, "Open a wide door, and make a smooth way for the produce of that Country to pass to our Markets before the trade may get into another channel." The "smooth way" would be a road through a formidable barrier, the Appalachian Mountains. The best route through that barrier was Nemacolin's Trail, a network of old Indian hunting paths that Washington knew well from former days.

As a visionary and master of military art, Washington believed the country's future depended on the existence of an overland route linking the cities of the East with the

rich resources of the territories of the West. His expedition set the course of America's first interstate, the National Road. It would take 50 years to convert Nemacolin's Trail into a federally funded highway, but its route would be the foundation on which America began the job of nation building. The 600-mile-long highway became a lifeline running through six states and connecting the Atlantic Ocean with the Mississippi Valley.

## BRUTAL TRAILS AND ROUGH ROADS

Most Americans, before the Revolutionary War, lived along the coast. Few dared to travel more than 100 miles inland. The interior of the New World was dense forest and rugged mountain, fraught with dangers and relatively unexplored. Only those native to the land, the Indians, understood the ways of this thick forest. For thousands of years they and their ancestors had followed the beaten-down paths of the wild beasts and buffalo that roamed the backcountry.

In 1751, Nemacolin, a Delaware Indian, joined with his friend Colonel Thomas Cresap to blaze one continuous trail along the many wilderness trails.

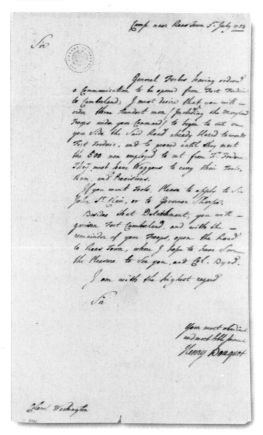

**This 1758 letter** confirms George Washington's involvement in the country's road building. He is asked to command his troops "to begin to cut on your Side the Said Road already blazed toward Fort Frederic..."

Cresap was an early entrepreneur, one of the founders of the Ohio Company, which had as its goal the settlement of lands on the Ohio River. He was an unusual man for his day, known among the Indians as Big Spoon for his endless generosity. He had a unique relationship with his native business partner, adopting Nemacolin's son years later.

The two-man team set out from what would become today's Cumberland, Maryland, in search of the best route west to Pennsylvania's Monongahela River. Scaling mountains and fording rivers, they hacked their way through the dense woods, notching trees with hatchets to mark a trail. When they were finished, they had created a rough way—bumpy, narrow, studded with boulders and stumps. But it was a way, nonetheless.

About a year later, George Washington, also a member of the Ohio Company, was one of the first to use this rugged path. The young and eager British officer had been selected by Governor Robert Dinwiddie of Virginia to deliver an eviction notice to the French army, which was staked out on land claimed by the British in the Ohio territory. Setting out on horseback along the trail marked by Cresap and Nemacolin, the 21-year-old major hit blinding snows and bitter temperatures as he made his way to the French encampment. When he finally reached the French, holed up in Fort Duquesne near today's Pittsburgh, they simply ignored his message, sparking the French and Indian War.

On the return trip to Cumberland, Washington's horses gave out and he was forced to dismount and return on foot. He deviated from the course of the trail, capturing the dangers of the land in his journal. Along the trek through the mountains, he recorded 20 Indian warriors en route to battle, seven scalped settlers, an angry Indian queen Alliquippa whom he mollified with a bottle of rum, and a near fatal plunge into icy waters while navigating a makeshift raft.

The British were determined to wrest control of America's west from the French, and Washington led the empire's early battles up and down Nemacolin's Trail. In the spring of 1754, he took control of Colonel Joshua Fry's detachment after Fry's death at Cumberland, the head of the trail. With a contingent of 60 men swinging axes and cutting a narrow swath for the advancing troops, Washington forged his way through the woods to meet the French. The French, however, ambushed Washington along the trail and defeated him. It was his first battle and it led to his only surrender, at Fort Necessity on July 4.

Desperate to remove the French from their stronghold at Fort Duquesne, the Crown sent Major General Edward Braddock to escalate the campaign. As an aide-de-camp to Braddock, Washington charged down Nemacolin's Trail once again, this time with 600 men chopping and cutting a 12-foot-wide road for more than 2000 British regulars, Virginia militia, and Indians. Supporting them were 150 Conestoga wagons and teams of horses dragging howitzers up the mountains. The terrain was so demanding that the Royal Navy loaned Braddock's force 30 seamen, who used ropes and blocks to lower gear down the precipitous slopes. But Braddock's Road, as it came to be known, was nearly impossible for an army to pass over. Steep grades and foot-high tree stumps slowed the troops, limiting their movements.

Washington found himself in the midst of another ambush. Before Braddock reached Fort Duquesne, the French and their Indian allies materialized out of the woods and cut them down. Washington survived, but Braddock was killed. With four bullet holes in his coat and two horses shot out from under him, Washington was again in retreat.

Nearly 50 years later, men working along the National Road's future path

**British General** Edward Braddock was ambushed by the French and their Indian allies, and buried beneath his own crude military road.

discovered General Braddock's remains. His body had been entombed in haste beneath his own road, the grave successfully disguised by wagon tracks and hoofprints to escape exhumation and desecration by the advancing enemy. It was a cruel beginning to a remarkable highway.

## THE ROAD THAT MADE THE NATION

"Let us bind the republic together with a perfect system of roads . . . let us conquer space!" was the battle cry of South Carolina statesman John C. Calhoun as he argued for funds for the National Road. Sounding more like an endorsement for NASA than the National Road, his words put into perspective just how farseeing an idea this road was. It was an experiment in transportation technology far ahead of its time.

But standing in the way of that idea was a very real problem: the Appalachian Mountains, a 1000-mile-long barrier that separated the bustling East Coast cities from the rich western lands. Rivers, the primary source of transportation at the time, were no help. The rivers ran in the wrong direction, north and south. The desired course of commerce was east and west. An overland route was needed, but these were so unusual for the times that they were called "artificial roads" or "dry roads." The real roads were wet roads—navigable courses along rivers, bays, and harbors.

**With the Louisiana Purchase,** the country's size more than doubled, expanding from the Mississippi River to the Rocky Mountains for a mere $11,250,000. With a vast tract of new land, the National Road became even more important to the nation.

*Treaty*

*Between the United States of America and the French Republic*

The President of the United States of America and the First Consul of the French Republic in the name of the French People desiring to remove all Sources of misunderstanding relative to objects of discussion mentioned in the Second and fifth articles of the Convention of the {8th Vendémiaire an 9 / 30 September 1800} relative to the rights claimed by the United States in virtue of the Treaty concluded at Madrid the 27 of October 1795 between

George Washington died in 1799, just a few years after he stepped down as President. He never saw his smooth passage to the west become a reality, but his vision led the country's third President, Thomas Jefferson, to take up his plan.

The time was right. Early in Jefferson's presidency, he and Napoleon finalized the purchase of the Louisiana Territory, extending the nation's borders dramatically west, to the Rocky Mountains. The country needed a "cement of union" to its new lands. The first phase would begin with the old Nemacolin's Trail in Cumberland, Maryland, and roughly follow its original path. The road would pass through western Pennsylvania, cover 131 miles to Wheeling in what is now West Virginia, and stop at the Ohio River. Obviously no one state could afford to build such a wide-ranging project, so the responsibility fell to the federal government.

In 1806, President Jefferson signed a congressional act establishing the National Road. It was a historic moment. Never before in the United States had the federal government built a highway, and never had a plan of this scale taken on such aggressive engineering goals. The law proclaimed the beginning of the federal government's participation in road construction.

Within months the President appointed three commissioners to select a route and plan the construction of the first federal interstate. Their instructions were clear but far from simple, especially since the road would be built by hand, one stone at a time. The path they marked must be as short as possible, serving as many people along the way as achievable. Its right-of-way had to be cleared to 60 feet, and the road could not have a grade higher than five degrees to the horizon. In other words, it must be as flat and straight as possible. As the commissioners' report to President Jefferson explained, they would "make the crooked ways straight and the rough ways smooth."

Setting off in September of that year, the commissioners and their team of surveyors, chainmen, and mappers scaled mountains and trudged through rivers, swamps, and fields. Their tools were simple: a 66-foot-long chain that was stretched for measurement and six-foot wooden stakes that were pounded into the ground to mark the road's future lines. The sun, compasses, and a table of logarithms helped in the survey. Spring floods threatened to wash away markers, so copious notes were taken of creeks, boulders, and other landmarks.

Along the way, the mappers boarded in local inns, where they sought advice from residents who knew the area and land well. They found the people excited about the new road. A stake in your front yard meant a way to get your produce to market and the annual passing of thousands of potential customers through your farm. You might even be hired to help build the road. It's no wonder that much of the land was donated and compensation for its taking was never considered.

In 1784, George Washington set out on horseback to survey a route for America's first federal interstate highway. He was concerned that American western settlers would become economically and politically dependent on "the Spaniards on their right or Great Britain on their left" if he didn't "open a wide door, and make a smooth way" for them to travel. That "smooth way" became known as the National Road.

Illinois

Indiana

Indianapolis

Richmond

River

Vandalia

Terre Haute

St. Louis

Mississippi

Wabash River

**President Jefferson** commissioned a team to select the route of the first National Road. They were to "make the crooked ways straight and the rough ways smooth" across the new nation, from the populated East to the wilderness in the West.

The work was difficult, and some of the mapping team thought the project was impossible at first. It took four years to chart and finalize the exact route. But the job they turned in was remarkable for its skill and forethought. Even modern road builders still admire the courage and sheer gall of those early planners.

The first section of construction began in the spring of 1811. The War of 1812 slowed progress until 1816, when work again picked up. England's hostile actions during the conflict were a vivid reminder of how important a passage through the disconnected country was. The road's construction was as much a military campaign as a commercial endeavor.

## A GIANT FIRST STEP

When the first crews gathered for the first day of work, they found a daunting obstacle waiting for them: Big Savage, a nearly half-mile-high mountain of Appalachian rock, situated right at the opening of the new road. It was a hellish start to their long construction project. The new road was meant to go over mountains—there was no other way west—but Big Savage was especially torturous. It pushed the limits of workers, beasts, pickaxes, and sledgehammers, as they chipped away at it inch by inch along a 14-mile stretch, eventually climbing 1000 feet. It's no wonder the first sections of the road were the most difficult and expensive to build.

As many as 1000 men at a time, most of them Irish and English immigrants, descended on the mountainous terrain between the Potomac and Ohio rivers, gouging a road into its jagged faces, creating a scene the locals would never forget. Choppers, grubbers, and burners worked with oxen to pull stumps from the roadway. Drivers used horse and wagons, and laborers pushed wheelbarrows as they dumped their loads of rocks, roots, and dirt over the ridges and into the ravines to level the grade. Axmen cleared the required 60-foot-wide right-of-way, chopping, cutting, and felling the trees. To clear the timber, laborers rolled the logs down the hills or piled them up and burned them on-site. Behind them were stump grubbers clearing muddy tree stumps from the roadbed, digging, cutting, and hacking away at the stubborn roots.

In the early work on the road, a 20-foot-wide, one-foot-deep roadbed was hand-dug and then filled with gravel hand-pounded to size with small hammers along a 131-mile swath through the mountains. At the bottom of the foot-deep foundation, workers first placed rocks no larger than seven inches, then layered a separate stratum of three-inch stones on top of them. In an early example of quality assurance, the hand-hewed stones were threaded through metal rings whose diameter matched the required size. Seven-inch stones for the foundation's base had to pass through seven-inch rings, and three-inch stones had to pass through three-inch rings—a tedious and laborious process.

**The road's original** construction was loosely copied from the French, who at the time were considered the finest road builders in the world. Ideally, but not always, the road was built with a base of seven-inch stone and a travel surface of three-inch stones.

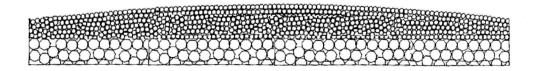

Two hundred years ago, a turnpike was any road that was paved with stones, wood or some combination of hard materials and required a fee for its use. After the toll was paid, an official would turn a pike (a long wooden pole acting as a gate), allowing a traveler to enter the roadway.

To deaden the boredom of pounding stones, competitions were sometimes held between stone breakers. Contestants with their trusty iron hammers with one-pound heads "as round as an apple" faced off, smashing away for hours as hard and quickly as they could. The grueling match lasted until one surrendered or time was called. Then their products were measured and a winner was declared. Interestingly, some of the stone breakers donned eye protection, an early form of safety glasses, made from steel and strapped around their heads with just a sliver of space for viewing. With their primitive eye protectors, they must have looked like fitful medieval knights swinging hammers instead of swords.

If the workers looked like knights, they didn't act like them. Fraud was a problem with the stone breaking. Families and individuals who contracted with the road builders were paid per pile of stones delivered. Too often the same piles were bought, sold, and bought again. To prevent the duplicitous act of selling and reselling the same piles of rocks, officials resorted to pouring whitewash over newly purchased stones to mark them as bought and paid for.

The last stage of construction was the road's surface. The crushed stones were covered with a layer of sand and gravel or clay depending on what was available in the area. The entire pile of rocks and sand was then compacted with a three-ton roller to harden the surface, leaving it "mettalled" and, most important, watertight. Ditchdiggers built culverts on both sides of the road to wick water away from it. Water was a particularly evil force in early road construction, as it is today. Winter thaws and summer downpours plague any road, but the National Road's steep embankments made it especially vulnerable. Extraordinary measures were taken to prevent hill runoff from penetrating the road's foundation. Unfortunately the ideal sealers, petroleum-based products like asphalt, were decades away.

As soon as a section of road was completed, it was turned over to eager travelers moving westward. Work teams were overrun by pioneers and farmers unwilling to wait for the road's completion. Inspired by dreams of western riches, settlers piled their wagons with their family's belongings and pushed past the end of the road. It was impossible to know if the highway was pushing the nation west or if the country was pulling the road along with it.

Whichever the case, the National Road's initial plans came to fruition in 1818 when the highway literally ran into the Ohio River, dead-ending at Wheeling, Virginia, which became West Virginia after the Civil War. George Washington's dream was realized and Thomas Jefferson's call to action had been heard. The nation's coastline and its expanding interior were at last connected. America's first interstate was a reality.

# A ROAD TO THE ROCKY MOUNTAINS!

That first part of the highway was the darling of the nation even though it came in behind schedule and over budget. The average cost of the mountainous road to Wheeling was a staggering $13,000 per mile, more than twice its original estimate of $6000 per mile. Nevertheless, it was crowded with travelers from the start and boasted an inn more than every mile along its route. Its construction through the mountains had been daily entertainment for the local people, as thrilling to watch as the road was to travel.

Excitement about the National Road facilitated the procurement of federal funding to extend the highway after a heated constitutional debate about state and federal powers was resolved. Speeches, proclamations, and much grandstanding filled the day on July 4, 1825, when construction began on the second half of the National Road. The new section would go through Ohio, Indiana, and arrive finally at Vandalia—Illinois's capital city at the time. One reckless prophet, a future state senator, saw it reaching all the way to the Rocky Mountains—or at the very least Missouri!

In Ohio, the mountains of western Pennsylvania and West Virginia give way to the rolling hills and flats of the Midwest, where glaciers long ago neutered the topography, shearing off mountaintops and dumping them into their own valleys. The land is level here, a good thing for the builders and financiers of the road. To help matters along, eager pioneers and farmers—too impatient to wait for a government road—had already cleared sections of the land, leaving only small clusters of trees and

**The Appalachian** Mountains stood directly in the path of the new road. Before Congress approved the construction of the first interstate over the mountainous area, the crumbling existing road made for some dangerous travel.

In **1879**, *Harper's Weekly* magazine hired a stagecoach and sent out a correspondent to recapture the early exuberant days of the National Road, when stagecoaches ruled and taverns were found as often as the milestones.

barren landscape in anticipation of the arrival of "America's Main Street."

But if the projected route was flat and already half cleared, its length was a killer. The first part of the National Road was 131 miles long. The second part was more than four times that. To keep costs low, legislators demanded that it be built straight as an arrow, linking state capitals and avoiding unnecessary meandering miles. The engineers stunned the public with their success in meeting those demands, making the road straighter and longer than any route ever seen before. It wasn't an easy job. Politically connected towns put up long but futile fights to direct the highway down their main streets in hopes of tapping the road's sources of revenue.

The Department of War was in charge of the National Road's construction west of Wheeling. The Army Corps of Engineers supervised the work and employed many West Point graduates as project superintendents. The officers applied the disciplines of civil engineering, road building, and bridge construction, acquired at the U.S. Military Academy. They raised the bar on the quality of construction, demanding that contractors employ the McAdam system, which was spreading throughout the world and would dominate construction and repair on the National Road into the next century. It was an expensive procedure, but the new interstate would be heavily traveled, and its earlier sections were already showing signs of wear.

John L. McAdam, a Scotsman, had a simple approach to building a durable highway: dig a shallow road foundation of 10 inches and fill it with smaller stones no more than two-and-a-half-inches wide and no less than one inch in diameter, "which shall unite by its own angles so as to form a solid, hard surface."

McAdam understood that water was a road's worst enemy. To keep water from seeping through a road's surface and into its foundation, he developed a nearly impermeable road surface using two-inch stones. Larger stones in the mix would eventually destroy the road's surface, pushing the smaller ones out of the way, allowing water in and eventually leading to the creation of a pothole. If a worker wasn't sure how to measure the stone, McAdam told them to use only the stones they could fit into their mouths. There were allegedly a few toothless road builders working on his roads.

**McAdam's road-** building technique, which improved the highway's resistance to water, was not used on the National Road until repairs were needed in the 1820s. Contractors had much leeway in how they built the original road, and there was little uniformity.

The McAdam plan proved that the weight of passing wagons and livestock would compact the stones into the road's surface, making it hard, relatively smooth, and therefore resistant to water. He stressed the need to keep the surface well maintained, allowing his shallow road to withstand as heavy a pounding as thicker and more expensive roads. Most important, McAdam believed in a road system's competent management. As an administrator, he stressed the need for supervision of the highway during and after its construction by well-paid professionals, an unusual concept at the time. The McAdam principles were so popular that a paved road is said to have been macadamized; and when tar is placed over a macadam surface, it becomes a tarmac surface. All of this quality was expensive, though. The final bill for America's first interstate came due upon its completion in 1839, a few miles short of the Mississippi River. After 33 years of surveying and construction of nearly 600 miles of highway and bridges, the final tab was $6,824,919—an enormous sum for the mid-1800s but not enough to build one mile on today's interstate system.

## SETTLING UP

If it weren't for the state of Ohio, there might not be a National Road. In 1803, Ohio had just become a state, the nation's seventeenth, and the federal government was selling Ohio's public lands to raise sorely needed revenue. The U.S. Congress decided that two percent of the profits from Ohio land sales would pay for the first federally funded interstate highway. This worked well during the construction stage; but with the road on its way to completion, the question of paying for its maintenance reared its troublesome head. Was it constitutional for the federal government to impose tolls along a road through the states to fund the highway's upkeep; and if so, did the government want to be in the business of running a turnpike?

No was the short answer to both questions. After much debate and constitutional hairsplitting, the federal government began transferring ownership of the National Road to the states years before its completion. That didn't happen until the last wrinkle was worked out—Maryland and Pennsylvania didn't want the highway. Its first sections, built decades earlier, were badly worn and in need of substantial improvements.

The final agreement was simple: The Federal government would pay for the improvements and install tollgates, then turn the road system over to the states. The states, upon inheriting the National Road, were responsible for its upkeep; but as an owner they could collect tolls from its users. The states created turnpike authorities to manage the road, and the Army Corps of Engineers erected tollhouses complete with a front porch, fireplace, many windows, and a list of

If you think paying tolls on today's highways is a pain, consider this. As late as 1913, Alabama required each of its citizens to work 10 days a year maintaining the state's highways. Affluent folk could buy their way out of the forced labor by paying a five dollar tax.

the rates nailed to the outside wall.

Once the tollhouses were built, the National Road became known as the National Pike. In the 1800s any road that was paved with stones, wood, or some combination of hard materials and charged a fee for its use was considered a turnpike. (Historically, a turnpike involved a spiked barrier or long pointed spear placed at a tollgate to prevent passage. To pay the toll and enter the roadway was to have the "pike" turned, or removed from the traveler's path.)

Traffic on the National Road was always heavy, and rarely gentle. Cows, sheep, pigs, and even turkeys walked east on the interstate as they were driven to market in droves up to 5000 strong. Horses, riders, wagons, coaches, men, women, and children all galloped, rumbled, walked, or limped along the highway. In the process, the road's surface took a constant beating, and the need for expensive repairs was constant. Toll collection became a serious business, although not always an appreciated one. One traveler complained on the record about an officious toll attendant who threatened to shoot his cow if he passed the gate without paying.

**Tollhouses appeared** when the states had to raise money for badly needed maintenance of the National Road. Fares were collected for horses, pigs, and anything else that passed along the highway, contributing to its wear and tear.

## STOGIES, HOGS, AND HEROES

Tolls, however, were not the only hardship along the road. It could take weeks to drive a herd to a market hundreds of miles away. In the winter, frozen and jagged road surfaces cut into the feet of the livestock, forcing unscheduled layovers in a friendly pasture while waiting for a thaw. A stray beast in the woods could take days to find, and securing a place to host a large group of cattle or

**It is believed** that Americans drive on the right side of the road because of the Conestoga wagon. Drivers who tired of walking alongside their wagons could pull out a wooden plank, called a lazy board, from the left side of the wagon and sit on it. Staying to the right allowed the drivers to remain seated without losing sight of wagons approaching on their left.

pigs for an overnight stay wasn't easy. Seeing an opportunity, some farmers along the highway leased their fields for a night of food and rest.

That was the wonderful thing about the road. Everywhere you looked, there was opportunity. It was the people's road, built by the settlers who lived along its path, on land they donated for its right-of-way. Women, children, farmers, and businessmen contracted to clear ditches, pound stones, and haul dirt to bring America's Main Street through their towns. After they built the road, they became its wagon drivers, stagecoach handlers, innkeepers, and toll takers. A man could start off earning $40 a week as a stone breaker, earn more than twice that wielding an ax and felling trees, go into business for himself driving a wagon or stagecoach, and then buy a tavern along the highway and retire a successful land speculator. In the heydays of the National Road, there were no limits.

The most frequent travelers along the road were poor farmers called movers. They came from the East in search of greener pastures in the West and turned the National Road into the main transportation artery to the frontier. Movers left their spent properties behind, packed up their pots and pans and everything else they could fit into a wagon or pushcart, and headed west. They were too poor to stay at inns and usually camped along the road at night. Mothers and daughters slept in the wagons, and fathers and sons on the ground, wrapped in blankets.

While traveling, these humble pilgrims were dwarfed by huge six-horse Conestoga wagons called freighters. Conestoga wagons, named after the

Conestoga River Valley in Pennsylvania where they originated, were the 18-wheeler trucks of America's first interstate. The wagons used for freight were over 20 feet in length and painted a Prussian blue with orange-red wooden wheels over five feet high. Their teams of horses were draped in brass bells and driven by rugged cigar-smoking men. These Conestoga wagons were serious business. They were built strictly for shipping freight and usually didn't have a single seat for a driver or passenger. There might be a lazy board on the left side for those times when the driver was too tired—or lazy—to walk.

These profitable long-haulers could handle up to five tons of goods. Westward, they carried coffee, sugar, and provisions to the pioneers. They returned east with wool, flour, hemp, and tobacco. Many drivers became rich moving freight on the new highway.

Conestoga wagoners on the National Road loved smoking cigars. Opportunity again. A small cigar maker in the town of Washington, Pennsylvania, produced an inexpensive and very popular Conestoga Cigar. The drivers of the big wagons snatched them up at four for a penny, enjoying them during their long hauls up and down the National Road. The official name, Conestoga Cigar, was lost forever when the drivers shortened Conestoga to stogie. The rough-and-ready drivers supposedly preferred the cheap roll-up stogie to an "alleged fine, high-priced cigar."

Stagecoaches moved faster and were painted brighter colors than the heavy Conestoga wagons. Coaches averaged about seven miles per hour, were upholstered with silk, and emblazoned with the names of the companies running them. Some names were as quirky as June Bug, Good Intent, and Defiance. Stagecoaches were the best way to travel, if you had the means. Nonetheless, it was a bumpy and bruising experience. Inside, passengers sat on three wooden boards. The man in charge sat on a board holding the reins and was usually a "sober and attentive driver." Luggage was strapped to the back, where it became covered with road dirt. Dust and mud were a given. There was never a trip without them.

In a matter of minutes a stagecoach could sound its horn, pull into a station, swap its four tired horses for new ones, exchange its first-class mail from a locked box and its second-class mail from a canvas bag, and be off again. Passengers were fortunate if the driver let them out of the carriage to stretch. When drivers made sharp turns at high speeds, they yelled down to their passengers, "Lean left!" or "Lean right!"

Going up hills was not a problem for a stagecoach; it was going down that

The National Road made the impossible a reality. As early as 1831, before the days of refrigeration, people in Ohio were able to eat fresh oysters that were shipped from the East by stagecoach.

In 1796, The Philadelphia and Lancaster Turnpike in Pennsylvania became the country's first financially successful turnpike. In 1940, the Pennsylvania Turnpike became the first modern turnpike.

was dangerous. At the top of a mountainous section of road, drivers might chop down a tree, tie it to the back of the coach and drag it down the road to slow their descent. If they could not find a suitable tree, they chopped branches into 10-foot poles, tying them with chains between the undercarriage and the rear wheels. The poles became a brake and forced the wheels to turn slowly. Drivers on the National Road used as many as twenty of these poles in a 100-mile stretch.

Steep hills and sharp turns were the least of the passengers' concerns. Robberies and accidents were always a threat. The trip was not for the weak, as the famous senator from Kentucky, Henry Clay, could have told you.

Senator Clay, whose statue was erected along the National Road because of his critical support in getting it built, was in a stagecoach accident in Uniontown, Pennsylvania. The Senator, after he was extracted from the overturned carriage unhurt, quickly spun the mishap, saying, "This is mixing the Clay of Kentucky with the limestone of Pennsylvania." The National Road was truly the Main Street of America and the country's heroes traveled it constantly.

For all its dangers and discomforts, there was a pioneer energy and optimism pulsing along the new road. It was called the King of Roads, the pride of the nation. Politicians eagerly took to its bumpy surfaces, meeting and greeting and pressing the flesh. On the King of Roads, a citizen could reach into a stagecoach and shake the hands of U.S. Presidents, past, present, and future. Citizens such as John Quincy Adams, Abraham Lincoln, and even Davy Crockett and Walt Whitman could be approached while eating at a roadside inn.

Andrew Jackson, Old Hickory, became America's first frontier President in

**America's tradition** of great bridge building began on the National Road with the completion of the Wheeling and Belmont Bridge. Its span of 1010 feet over the Ohio River was a record, the longest in the world.

1829. He was one of the National Road's biggest supporters. Jackson was born in a backwoods settlement, with little formal education. He engaged in brawls, killed a man in a duel who insulted his wife, and was a victorious general in New Orleans during the War of 1812.

At this point in history, Americans were in love with their new road and their new country. They declared their patriotism in the names they gave to the inns that lined the road: the American, the National, the Constitution, the Washington, the Ben Franklin, and the General Jackson. Inside those inns, locals gathered in the public rooms to listen to gossip and news, trade ideas, even sing and dance. Energy flowed along each mile as an unprecedented exchange of information, goods, and services passed from one American to another. Travelers stopping at an inn along the National Road often said it was time to "water the horses and brandy the men." At the end of each day, herders and wagon drivers slept on barroom floors, and the young nation's illustrious statesmen and heroes clambered upstairs to get ready for a new day. It was a remarkable life along America's first interstate 150 years ago.

## THE KING'S CROWN

By the mid-1800s, America was thinking and acting big, and its new roadway was its stage. In 1847 work began on the National Road's crown, the world's longest suspension bridge. Before there was a Brooklyn Bridge or a Golden Gate, there was the Wheeling Suspension Bridge. The immense structure suspended in the air was considered a noble monument, not just a piece of highway.

The new bridge spanned an unthinkable 1010 feet between bridge towers, was 92 feet high, and its bridge deck was wider than the highway on land. It cost nearly a quarter of a million dollars to build and was the greatest structure of its kind in the world. Designed by Charles Ellet, it could support 297 tons, a stupendous achievement at the time. In 1849, the Wheeling *Daily Gazette* put it in understandable terms, explaining that the bridge could safely hold a total of 32 laden wagons, 172 horses, and 500 people, or an army of 4000 men.

Not everyone was excited about the bridge. Pittsburgh, a rival frontier city up the Ohio River, was concerned that it would put them at a disadvantage if it blocked the passage of steamboats, an important commercial link for that city. The National Road had already established Wheeling as the country's gateway to the western United States. Now its bridge was seen as a threat to Pittsburgh's development, a physical barrier to commerce. The city retaliated.

Pittsburgh manufacturers placed massive, 80-foot-high smokestacks on a fleet of six boats and floated them down river. Smokestacks of the day were no higher than 60 feet, so the intention was clear—to force a collision with the bridge and declare it an obstacle to navigation. One or two of the boats suffered damage, and the bridge workers cut off the top of at least one of the smokestacks. Pennsylvania, representing Pittsburgh, sued the builders of the bridge. The state won in court, but the victory was overruled by Congress, which declared the bridge a post road under its control, with every right to be exactly where it was.

But for all the energy and political clout behind it, one thing the road couldn't do was stop progress on other fronts. A newfangled invention called the steam engine was making its presence felt, giving new power to the infant railroad industry. Snaking through the mountains and along the frontier, the "iron horses" of the railways had a speed and stamina no flesh-and-blood pony could match. The days of the horse and wagon were numbered.

In 1852 the first train chugged into Wheeling from Baltimore, and a cloud began to gather over the King of Roads. Before long, it fell into a period of decline, many thinking a death knell had been sounded. No one understood that the National Road's true glory days were ahead.

*Opposite:*
**The first bridge** was torn apart in a windstorm in 1854 but was rebuilt immediately. Still in use today, the Wheeling and Belmont is the oldest suspension bridge in the United States built as a roadway and is a National Historic Civil Engineering Landmark.

The term Parkway was first used by landscape designer Frederick Law Olmsted, in the mid 1800s, when he introduced his idea of a new tree-lined boulevard that served as both a road and a park. The first Parkway led to and from Prospect Park in Brooklyn.

## CHAPTER THREE

# Roads of Iron and Mud

**O**N MAY 10, 1869, on a 5000-foot-high summit near the Great Salt Lake, a golden spike struck by a silver-headed sledgehammer was driven into the last railroad tie, completing the Central Pacific and the Union Pacific's transcontinental railroad. The spike and hammer were rigged with telegraphic wires so America could listen to the ceremonious completion of the first railroad in the world to span a continent. Reporters compared the resounding clash of the silver hammer and golden spike to the shot heard round the world at Concord and Lexington. America was at last physically united by a single road of iron.

A telegram of congratulations was dispatched to President Ulysses Grant; the Liberty Bell rang in Philadelphia; and cannons blasted away in Washington, D.C., and San Francisco to celebrate the momentous occasion. The nation was expanding, the West was being opened, and the railroads ruled the day.

It is unlikely anyone foresaw the limitations of the railroad during these initial celebrations. It would be half a century before Americans would begin to chafe at the limited movement available to an individual riding on the roads of iron rails. Over the next half century, the iron horse was the chief form of mass transportation and Americans struggled with the railroad's dominance over their lives. Individual travel schedules, shipments of goods, and the cost of shipping those goods were in the nearly total control of the railroad executives and their partners. The nation

*Opposite:*
It was a glorious day—a young nation united with a road of iron. But railroad travel would soon prove too restrictive for restless Americans.

35

had no choice but to depend on the railroads. Nearly every mile of road in the country was a sorry stretch of mud and bumps, uncomfortable in the best of times, impassable in the worst.

The United States, however, was growing into an economic and militaristic world power. The country was restless and impatient at the turn of the nineteenth century, longing for an open road and the ability to move freely up and down it.

## BICYCLE MANIA

The first highway lobbyists in the United States were a group of bicyclists, The League of American Wheelmen (L.A.W.) organized in 1880. Ironically, today bicyclists are not allowed on the Interstate System in most states.

In the last half of the nineteenth century, the self-propelled individual traveled the earth at the same breakneck speed that Cleopatra or Aristotle had, about three miles per hour. Unless you were on a ship, train, or horse, your mobility was severely limited. That is, unless you had a contraption that was gaining popularity called a bicycle. Cycling was the first modern phenomenon that fulfilled the American public's craving for the freedom to move anywhere at will. It set the stage for the automobile.

It took a while to get it right. The old-style bicycles with a very high front wheel and small rear wheel that arrived on the scene in the 1870s required the adventurous rider to sit on top of a wheel that was commonly 58 inches in diameter, putting the rider's head a dangerous eight feet above the road. These bikes, with no springs and with solid steel or wooden wheels, made for a bone-rattling ride and a serious fall from grace if the rider lost his balance.

The next step was the one that put the masses of Americans on wheels. It was the "safety bicycle." With two rubber air-filled tires of the same size, a cushy seat, and suspension springs, it allowed riders to bike along in safety and comfort. The safety bicycle fueled a craze for movement that consumed America. Bicyclists organized themselves into clubs, and then organized day trips and weekend-long "century trips"—100-mile rides along the country's bumpy roads. Sunday preachers began to complain about empty pews at Sunday services, and barbers complained about people missing their Saturday haircuts. Too many people had taken to the road.

The bike became a dynamic influence in America's social and political circles. Men and women of different classes intermixed in the biking clubs, and no less a personage than the suffragist Susan B. Anthony endorsed them. "Let me tell you what I think about bicycling," she proclaimed. "I think it has done more to emancipate women than anything else in the world. I stand and rejoice every time I see a woman ride by on a wheel."

Never mind that the woman was apt to be wearing scandalous bloomers, dark cloth leggings, and—most shocking—shorter skirts, all to keep her garments from entangling in the bicycle's chains.

**The whole family** climbed onto bicycles to explore the world beyond a town's limits. Early automobile design owed much to the bicycle.

1899 :: Model :: 1899

Hawthorne Bicycles
$18.00 Net

GENTLEMEN'S HAWTHORNE—1899 Model
Color, Dark Myrtle Green

GUARANTEE FOR 1899 MODEL HAWTHORNES—Our guarantee is absolute protection. We warrant every Hawthorne Bicycle for one year, and agree with every purchaser of a Hawthorne to make good, either by repair or replacement, any defect in material or work, and not a result of misuse or neglect. We further agree that upon any part of a Hawthorne bicycle is except tires) wear to us, charges prepaid, we agree that, any where claim for defect is allowed, we will reimburse the owner of such defective part for the amount of transportation charges paid, and return perfect part once or cost to our customer

SPECIFICATIONS B 1084—Name: Hawthorne. Cranks: 6½-inch diamond
(Indianapolis B block, best quality), strain center and shape). Chain: Standard, 5/16 hardened centers and rivets
tion 22 or 20-inch. Finish: Dark myrtle green, nearly hand-sided. Frame: Regular 24-inch, op
78, 10-inch. 26-inch. Pedals: Bridgeport, rat trap. Gear: Regular 75 equals
Adjustable. Pedals: 1 Bridgeport sprockets are used on 72 gear. 10 and 28. Handle Bars:
Extebure Needle Co.'s best No 2. 22 front, 36 front. Tires: 1⅜-inch, padded top. Spokes: Tangent,
tube. Tool Bag: Containing wrench, oiler, repair outfit and spanner. Tread: 4½-inch. Tubing:
Shelby cold-drawn seamless. Wheel Base: 42½-inch. Wheels: 28-inch. Weight: About 25 lbs.
Please inform us, when ordering, at all our bicycles are made and stored during the winter months,
ratated and to paper bags, (track or prompt shipment). Turn one of the reasons why we are able to base so low a price.

Read What We Say on the Other Side of this Paper

Montgomery Ward & Co., Michigan Avenue
Originators of the Catalogue Business and Madison Street Chicago

By the 1890s, bicyclists were traveling all over the country. Organized into a 100,000-member-strong organization—a huge number for the day—they called themselves the League of American Wheelmen. And they were the loudest single voice advocating better roads.

But as important as the bicycle was in putting America on the road, it was only the fuse that powered the biggest explosion in American mobility—the automobile.

## BIRTH OF THE AUTOMOBILE

The first American-made automobile powered by internal combustion and fueled by gasoline was born from the efforts of two bicycle mechanics. In Springfield, Massachusetts, Charles Duryea, 31 years old, and his little brother Frank, 23, worked together to develop the first gasoline automobile. They started with a broken-down horse carriage with wooden wheels that Charles bought for $70. Charles and Frank rigged the buggy with a crude one-cylinder gasoline engine and had it running up and down the streets of Springfield by 1893. Unfortunately, it had no brakes. To stop the thing, they had to run it up against a curb.

The Duryeas called their early vehicle a Buggyaut—part buggy, part automobile. The brothers went to work on an improved model of the

Buggyaut, this time with brakes, for America's first automobile race in 1895, the *Times-Herald* contest in Chicago, Illinois.

The Duryeas won the race handily, beating out several other autos, including one of the leading driving machines of the day, the European-made Benz.

The well-hyped contest brought national recognition to the automobile as a serious mode of transportation and to the Duryea brothers as America's first manufacturers of a marketable gasoline car. The bicycle had turned into an automobile by way of the horse and buggy.

At this crossroad in transportation, the cycling world's innovations were being absorbed by the infant automobile industries. Designs born from bicycles were used to drive the automobile faster and more smoothly and to allow it to brake harder with better control. Differential gears, rack-and-pinion steering, band-brakes, and pneumatic tires—all invented by bicycle manufacturers—were employed by the automakers. Even the bending and plying of metals into mechanical parts was a credit to the bicycle industry.

But there was an important difference between the bicycle and the automobile. It was that loud, smoke-belching but potentially powerful gas engine that put the auto in automobile. No pedal power needed. At the same time the Duryea brothers were driving their Buggyaut, Henry Ford was experimenting with his own gas engine in his kitchen, eventually attaching it to a 500-pound carriage equipped with bicycle tires. Other American mechanics and visionaries were fiddling with the new technology, trying to harness the engine's power and put America on another kind of wheel. To do that, they would also need roads.

## SPECIAL AGENT NUMBER ONE

Tucked into the Agricultural Appropriation Bill of 1894 was $10,000 for a yearlong study and dissemination of information on national road conditions and road-making techniques. That small stipend ultimately led to the creation of the Office of Road Inquiry, the country's first federal road agency. It was a one-year trial run. The mission: to "make inquiries in regard to the systems of road management throughout the United States." The office was housed at the Department of Agriculture, as highways were seen as a farm issue. "Get the farmer out of the mud" was the mantra—connect the farmer to his markets with good roads.

General Roy Stone, a good-roads activist and member of the League of American Wheelmen, became the new office's first special agent. Energetic and ambitious to succeed with his mission, he was an ideal choice. He arrived armed with a degree in engineering and badges of courage from the Civil War, where he was badly wounded on the first day of the Battle of Gettysburg while

*Opposite:*
**Charles Duryea,** a former bicycle mechanic, proudly steers one of America's first automobiles. He and his brother Frank performed test drives at night in Springfield, Massachusetts, working in the dark, to avoid embarrassment if the vehicle failed.

The first road in the United States to use asphalt pavements was New York's Fifth Avenue, in 1872.

*39*

*Above:* **In 1897 the** first section of federal test road is built in New Brunswick, New Jersey.

*Below:* **General Stone** enjoying a cigar. He was the first chief of the country's first federal road agency.

holding McPherson's Ridge, a crucial position for the Union Army. The hard-charging general turned his headquarters at the Office of Road Inquiry, located in the department of Agriculture's attic, into a war room. He created a large map of the United States and began charting the country's highways on it. At this point, there wasn't a single accurate highway map of the United States. Not one person knew how many miles of highways the country had or where they lay.

The general and his secretary and the third and only other employee sent out letters to every county seat in every state requesting their assistance in charting their roads. They asked the states to indicate the condition and material makeup of each of their roads and include which were made with hard surfaces and which were just dirt. It took until 1904 to complete that first federal survey, and it wasn't until 1907 that the results were tabulated and made public.

General Stone also set out to single-handedly addict the nation to good roads. Practical and strategic, he took his measly $10,000 annual budget and planted very short sections of well-built roadways around the country. With the help of road-building manufacturers who donated equipment, federal employees built the roads in well-publicized demonstrations. These object-lesson roads were only hundreds of feet long, but the general knew that once people traveled across a hard, smooth road there would be no going back to muddy and impassable trails. It was his own version of seeing is believing. The first object-lesson road was built in Atlanta, Georgia, in 1896 at the National Good Roads Parliament and was just a quarter of a mile long. The following year the general built one in New Jersey, and within three years he completed 21 object-lesson roads in nine states. There were more demands for his short roads than he could build.

## THE ROAD SHOWS

The object-lesson roads continued even after Stone left office. Good Roads Trains—rolling trade shows moving with considerable pomp—spread the good word about good roads. The railroads, lobbying for better roads to carry freight and passengers to their stations, supplied the cars and locomotives. Manufacturers of road-building machines supplied the building equipment, and the Office of Road Inquiry supplied the experts, fully equipped with speeches and a lot of advice. Well publicized, the trains were met with large and curious crowds all along their routes.

"An itinerant college on wheels has come among us," said U.S. Senator J. W. Daniel of Virginia after a Good Roads Train pulled into Lynchburg, Virginia, "This college does not teach out of books, nor solely by word of mouth. It teaches by the greater power of example." Once he had said his piece, the train unloaded its "titanic" machines and summarily built a "good road over an old and bad road" to the fascination of the crowd.

This was the most elaborate of the popular road shows, costing $80,000, a huge sum for the times. The train was sponsored by the Southern Railway and began its tour on October 29, 1901, in Alexandria, Virginia, traveling for five months and 4037 miles. Others traversed the country from Seattle, Washington, to Buffalo, New York, building demonstration roads and spreading the gospel of Good Roads.

Teddy Roosevelt was the first President of the United States to ride in a car. Warren Harding was the first President to drive a car.

*Good Roads Trains—rolling trade shows moving with considerable pomp—spread the good word about good roads.*

**The Good Roads Trains** resembled a traveling circus. They carried equipment and sleeping cars for the engineers on board who built one-mile strips of "good roads" at stops along the way. Thousands gathered in amazement to watch.

**Dr. Jackson at the** wheel of his country-crossing car. Next to him is his mechanic and former driving teacher, Sewall Crocker. In front is Bud, the faithful watchdog.

*Opposite:*
**Cross-country travel** was a hazardous business, but the breakdowns gave Bud a chance to explore the territory.

## THE GREAT AMERICAN ROAD TRIP

On May 18, 1903, in the prestigious San Francisco University Club, a group of men in the bar debated if it was possible to cross the continental United States by automobile. It had never been done, and they concluded it was still impossible. Drawn into the conversation was a 31-year-old doctor from Burlington, Vermont, who was wintering in California with his wife. Dr. H. Nelson Jackson had learned to drive months earlier and had no intention of risking his life to drive across the country. That was until the group in the bar bet him $50 that he couldn't travel by automobile between San Francisco and New York City in less than 90 days. Jackson spontaneously agreed to the challenge, and the Great American Road Trip was born.

After he broke the news to Mrs. Jackson and made plans to send her back east by train, the good doctor rounded up his driving instructor, Sewall Crocker, a 21-year-old mechanic and chauffeur. Crocker was authorized to acquire a sturdy automobile, while Dr. Jackson procured the critical supplies. These included a shotgun, a rifle, and two handguns for protection against road

Thomas Stevens, a Colorado coal miner, was the first to ride a bicycle across the United States. He left Oakland, California in April, 1884, and arrived at the Atlantic Ocean in August.

bandits; fishing gear, pots, pans, and sleeping bags since they would not see restaurants, hotels, or any sign of human life for days at a time; an ax, spade, ropes, and pulleys for pulling the car out of the inevitable mudholes; and extra tools, gas, and water containers since the modern gas station didn't exist. They decided to make up their course as they went along, relying on a compass, the simple maps that then existed, and the goodwill of the people along the way.

On May 23, Dr. Jackson and Mr. Crocker quietly departed San Francisco and headed for New York City in a 20-horsepower, two-cylinder car built by the Winton Motor Carriage Company of Cleveland, Ohio, a reliable and well-built car of the time. Along the route, there were no street signs or a single marked cross-country highway route to follow. The traffic signal, stop sign, and painted line in the middle of the road had not yet been introduced in America.

Dr. Jackson named the car the *Vermont*, after his home state and ultimate destination. The tiny wooden-and-metal car was driven forward by chains attached to the rear wheels, propelling it to a top speed of 25 miles an hour. The skinny and weak tires blew out with regularity. Each pop of a tire meant peeling it off its wooden rim, patching the inner tube, and then hand-pumping it with air. Treads for tires had not yet been invented, so the duo wrapped the rear wheels with rags and ropes to create some traction for climbing and braking.

Avoiding the deserts of the southeastern United States that had defeated other attempts at a transcontinental motorcar crossing, Dr. Jackson and Crocker headed north and east. Somehow they managed to average 71 miles per day on just two sets of tires. It took a week for them to get out of California. They avoided the snow-blocked trails of the Sierra Nevada as they moved through Oregon and Idaho, where they picked up a bull terrier, rescuing him from a career of dog fighting. They called him Bud.

*The little band of three with their driving goggles—Bud had his own pair—frightened many a passerby who had never seen an automobile or a goggle-clad dog before.*

The trio arrived in Omaha, Nebraska, on July 12. The Rocky Mountains and the Continental Divide were now behind them and only two weeks of travel were left in their odyssey.

At this point, running out of gas had forced Crocker into a nearly 30-mile hike and meant an unexpected night alone under the

45

*Opposite:*
**Mud kept everyone** from moving, whether on two wheels or four. Early cars like the one pictured had light bodies and a high clearance between car and ground in order to navigate deeply rutted roads. The car of the times fit the road of the times.

Horatio S. Earle founded the American Road Makers in 1902. Every state's capital would be united with an interstate system ARM called the Capital Connecting Government Highway. Membership was five dollars. Horatio said, "The membership will be five dollars. This is not a cheap affair, and cheap members will not be solicited."

stars for Dr. Jackson as he stayed with the *Vermont*. Many a mudhole had laid claim to the car along the way, requiring the blocks and pulleys or, in severe cases, a team of horses to pull it free. When the travelers ran into covered wagon trains along their journey, the *Vermont* was forced to back down the mountain trails and let them pass. In 1903 the western United States was still untamed land, and Arizona, New Mexico, and Oklahoma were territories waiting for statehood.

By now, word of their journey had spread, and the *Vermont* was met with crowds and fanfare as the traveling trio rolled through the Midwest and into the eastern United States. They were being chased, however, by two other teams that had left California in an attempt to be the first to travel—as one newspaper put it—"Ocean to Ocean in an Automobile Car." One team was driving a Packard and the other an Oldsmobile. Both were hoping Dr. Jackson would fail.

The other teams must have been disappointed when they learned that the road-weary crew chugged down New York City's Fifth Avenue in the mud-baked *Vermont* at four in the morning on July 26, 1903. Dr. Jackson won his $50 bet but he estimated it cost $8000 in wear, tear, and expenses to collect on the honor.

And when the good doctor arrived in his home state, he hardly got a hero's welcome. As a news dispatch from October 3, 1903, explains, "Dr. H. Nelson Jackson, first man to cross the continent in an automobile, was arrested in Burlington, Vermont and fined for driving the machine more than six miles an hour."

## MUD AND DEVIL WAGONS

When Dr. Jackson, Mr. Crocker, and Bud were bouncing across the continent, there were almost no paved highways suitable for an automobile. There was, however, plenty of mud. Mud was everywhere 100 years ago. When nearly every road was a simple dirt path, mud controlled people's lives, limiting mobility and dictating travel schedules. It increased the cost of living by choking the main arteries of commerce, making the delivery of goods and services an arduous and unpredictable event.

In the rural areas, winter storms and spring thaws created hellish muddy seasons, keeping children from school, doctors from house calls and the mail from collection. Mud held the nation prisoner, keeping most people from traveling more than 20 miles from their homes. Horses had been known to drown in mudholes before they could be unhitched from a wagon. Entire towns were marooned for up to nine months a year while they waited for road conditions to change from impassable quagmires of slop to barely navigable byways.

In the cities it wasn't much better; only the most heavily traveled commer-

**In 1911, the first** transcontinental trip by truck was a long one, slowed down by obstacles like railroad tracks. Traveling at an average speed of about three miles per hour, the *Pioneer Freighter* took nearly three months to cross the country.

cial streets were paved with cobblestones or bricks. Mud restricted movement on nearly every urban boulevard and avenue in America, making the simplest intercity travel a messy chore. The lack of good roads restricted the urbanite and lowered the quality of life. Bad roads kept Americans from the good times, and mud was to blame.

The federal government's road census exposed the sad state of the nation's highway system. America, it was revealed, had only 141 miles of paved rural roads in the entire country, of which only 18 miles were covered with a bituminous blacktop. Two of those miles were in Massachusetts, and the rest were in Ohio. Only seven percent of the total 2,151,570 miles of rural roads in the United States had any hard surface on them. This meant a token cover of gravel, stones, planks, seashells, clay, or any reasonably durable material that could be found locally. The rest of the two-million-plus miles of road were dirt, dust, and mud.

These primitive roads were regularly torn apart by horse and cattle hoofs and undermined by constant water erosion. Thin wagon wheels sliced through them, creating mile-long ruts. Automobiles added to the problem as their excessive weight and soft rubber tires forced the mortar out from between the stones.

Maintenance in rural America was somewhere between unnoticeable and nonexistent. Roadwork was the burden of those who lived along the road's edge. If these good citizens bothered at all to carry out their chores, they rarely did more than shovel dirt onto the highway and wait for passing horses, wagons, and automobiles to pack it down. Even the most responsible laymen knew little or nothing about highway engineering and what was required for a lasting fix.

But the country was beginning to change, thanks in part to the example set by the famous transcontinental trio and others like them. Americans followed their lead and ventured out of the cities and into the countryside with their cars. Motor vehicle registrations exploded in the United States between 1900 and 1910, even though the roads remained treacherous. In the first 10 years of the twentieth century, there was a 5500 percent increase in motor vehicle owner-ship nationwide.

In 1902, the newly formed American Automobile Association (AAA) took over the role of the League of American Wheelmen as people abandoned their bicycles for automobiles. AAA organized their trips, printed maps, and placed signposts so members could find their way into and back out of rural America. Farmers cursed the "devil wagons" as they roared—by early-century stan-dards—down the rural roads, kicking up dust and exciting the animals. Never mind. The country had fallen in love with the automobile.

## AMERICA'S INTERSTATE TRUCKING

There was a dark side to this automobile world, and it was called the truck.

In the early 1900s, the truck was loathed nearly as much as the automobile was loved. Autos were about pleasure. The truck was about work and a hard liv-ing. Trucks were slow and weak, and they blocked traffic on the narrow city streets. Relegated to the short shipping hauls that the railroads couldn't be both-ered with, trucks performed local deliveries of mostly household items.

But events in Europe were changing the way the world viewed roads and the motor truck. World War I was underway in France by 1914, and the truck was about to come into its own. In the Battle of Verdun, the longest and one of the bloodiest of the war, the French put 8000 trucks into action as they had never been before, carrying reinforcements, ammunition, and supplies to the battlefront and ferrying the injured back.

Europe had long been ahead of America in its road building, and France, because of its advanced highways, was able to dispatch the trucks at an incred-ible rate of one every 25 seconds. The roadways they used became a lifeline for the men trapped in the trenches. The Sacred Path it was called, as a testament to its significance. It still carries the name today. Thousands of trucks and the Sacred Path enabled the French to resist one of the most intense German attacks of the war.

The United States declared war on Germany on April 6, 1917. American industries prepared as President Wilson, reluctantly leading his nation into war, called for complete mobilization. Within the year, the President had taken con-trol of the national railroad system to bring order to the flow of goods for the

Harvard College in Cambridge, Massachusetts, was the first in the United States to offer advanced classes in the science of highway engineering.

**In 1917, trucks were** stronger than most bridges. Goodyear's Wingfoot Express crashed through a wooden bridge on its way from Boston to San Francisco. The trip was part delivery run but mostly a promotional tour for Goodyear's new pneumatic tires.

war effort. Supplies and munitions headed for Europe took precedence over all other railroad cargo. Basic necessities like milk and fresh produce were summarily removed from shipping schedules, causing a national crisis. The country had become so dependent on a single mode of transportation, the rail system, that it was unable to find alternative ways to ship goods and materials.

Yankee ingenuity and innovation arrived in the form of the truck.

Suddenly it became patriotic to "Ship by Truck" as one national campaign urged. Farmers, desperate to get their crops to market before they rotted in their fields, loaded up their farm vehicles and drove them into the city, farther than they ever had before, to reach their markets. Since many farmers didn't have their own trucks, merchants drove their fleet vehicles to them, sending their drivers out of the cities and into the country for the first time. Almost instantly, interstate trucking became an industry, as routes opened between Philadelphia and New York and between Toledo and Detroit. The trucks not only successfully picked up the overflow from the railroads, but they streamlined the overall shipping process. Eliminating the railroads meant merchandise was traveling directly from farm to market, reducing breakage during shipment and lost time while stored at railroad warehouses. Trucks performed better on the short hauls than the trains could ever expect to. As a result, the new business the truckers took from the railroads would remain theirs for good.

On April 9, 1917, three days after war was declared, the Goodyear Tire Company launched the Wingfoot Express, a nonstop 24-hour service between Akron, Ohio, and Boston, Massachusetts. The first truck rolled out of Akron with a load of newly minted tires and headed for Boston. Driving through heavy spring rains, the trucks completed the run 21 days and 28 tires later. The return trip was reduced to seven days because of better weather and dryer roads.

With enormous profits to be gained from tire sales if interstate trucking took hold, Goodyear launched the country's first transcontinental truck run between Boston and San Francisco in September of 1918. The patriotic inaugural run departed Boston, Massachusetts, with a pilot car guiding two Wingfoot Express trucks loaded with cotton-cord fabric for tire fabrication in Akron, Ohio. In Ohio the material was exchanged for new airplane tires that were delivered to San Francisco, California. The 3700-mile trip, 70 percent of it over muddy roads, lasted just 12 days even though the trucks fell through several rotten wooden bridges. Goodyear's pneumatic tire increased the speed of the trucks and spread their weight over the highway, causing less damage to its surface than solid rubber or steel wheels. In order to keep the trucks moving through the night, the Wingfoot Express trucks employed two drivers for each vehicle, which were designed with the first-ever sleeping cabs, a cramped space behind the operator's head. The Wingfoot Express trucks became a rolling billboard for American interstate trucking and a call for better highways.

**Trucks came into** their own during World War I, delivering crucial supplies at home and on the battlefields of Europe.

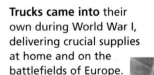

## DETROIT TO BERLIN

In September of 1917, the United States Army ordered 30,000 trucks from America's automobile man- ufacturers, most of them located in Detroit, Michigan. The country was fighting the war in Europe, and trucks were at the top of the list of equipment needed to sol- dier on. The problem was getting them from the Midwest to the East Coast, where ships were waiting to deliver them to France.

In times of peace the railroads would have been the obvious choice; but with the war's other demands, they were unable to handle a delivery of this magnitude. The army's Quartermaster General ordered the 30,000 trucks to be driven under their own power from factory to ship.

There was a problem however. Trucks for the most part were still only being used for short hauls. Despite Goodyear's and other companies' promotional tours, it was unthinkable to rely on trucks for long hauls. Besides, no one could look at a map and plot a course with any certainty. There were no federal highways marked with numbers. No one in the government could be certain of the condition of the roads that were on the maps, or if those roads were even still in existence. The federal government's survey was dated, and little new information reliable enough to pin the mission on was available.

On November 22, 1917, a reconnaissance mission was dispatched from Toledo, Ohio. The objective was to find a series of dirt roads linking Detroit's factories and Philadelphia's shipyards. The crisis became an important lesson in mobilization. It was now clear that the national defense depended on a strong highway system.

Three weeks after the primary route was selected, the war-bound trucks took to the road carrying banners reading DETROIT TO BERLIN. Running southeast

through Ohio to Pittsburgh, the path ascended into the Allegheny Mountains of Pennsylvania before reaching Philadelphia. Secondary routes were developed along the old National Road to the port of Baltimore and another through Cleveland and Buffalo and Albany to New York City. In one of the worst winters ever, convoys of trucks began rolling down the main streets of small farming towns. Winter storms blew six-foot-high snowdrifts across the mountain passes, requiring 24-hour vigilance of highway crews to keep the trucks moving. Each truck was loaded with up to five tons of ammunition and spare parts and traveled at a top speed of 14 miles per hour on its solid wheels. It's a wonder they ever made it to the battlefields.

Victory came at the cost of the country's road network. The heavy trucks destroyed nearly every mile of major roadway in the United States. This was not a case of wear and tear. It was, in the words of one federal official, the "simultaneous destruction of the entire road system." Almost overnight the thinly surfaced, narrow roads were wiped out by the brutal steel and solid rubber wheels of the heavy and primitive trucks. Every type of surface—brick, stone, gravel and clay—failed under their weight. An entire highway in Delaware was ruined by the passing of a single overloaded truck.

In Michigan, however, a stretch of highway built with an extra thick concrete surface survived well under the heaviest traffic the war effort produced. There was hope for the roads of the future.

*Victory came at the cost of the country's entire road network. The heavy trucks destroyed nearly every mile of major roadway in the United States.*

One of the first motorized road-grading contraptions, succeeding the old split-log drag, which was pulled by a team of horses. The weight of the man standing on the drag helped flatten the road.

CHAPTER FOUR

# The Great American Partnership

**T**HE INTERSTATE SYSTEM is a wonder of engineering and construction. But the partnership that built it might be even more impressive than the concrete-and-steel highways that stretch across our nation.

The partnership that created the Interstate System is a voluntary arrangement between the federal government and the individual states. At any time, any one of the states can take down the red-white-and-blue shields that have become so familiar along our highways and quit the road-building program.

They don't, of course. There is no sound reason for any state to opt out of the federal program, which is the core reason for its success. It has created a bond that is uniquely beneficial to each member. Like a good marriage, it has survived challenges that would have destroyed a weaker commitment.

It wasn't always that way. In the early days of road building, the country struggled with the issues of who was to pay for the construction and maintenance of the roads. Clearly, the job had turned out to be too much for the individual counties, and even the states. Nationwide, small and poor counties struggled with the responsibility for the simple roads then in existence. Building and maintaining thousands of miles of roads, capable of standing up to the growing number of automobiles and that road-chewing machine, the truck, was beyond local reach.

And then there were the tricky questions of how and where the roads would connect from state to state. How would that be decided?

The federal government was the obvious choice to resolve the issues and to pitch in with the funding for the roads. But Americans, democratic to the core, insisted on strong control of their local affairs, and building roads was one of them.

## AN IMPORTANT PAGE

The significance of Logan Waller Page was not lost on him. Page, a successor to General Stone, was confident to the point of righteousness. He used his scientific mind, as well as his powerful family and friends, to pave the way for the delivery of good roads. He did more for the country's long-term road building agenda and the development of the Interstate System than he was able to do for the Good Roads causes. His short-term agenda ended with his sudden death in 1918.

Born into America's aristocracy, and part of the progressive movement that combined education and a scientific approach to solving social ills, Page set out to give Americans the roads he thought were best for them, built by the people he approved of. There were "a lot of human vultures feasting on the road movement," he thundered, and "the cheap charlatanism of the professional promoter and the bungling efforts of the well-meaning but uninformed citizen should no longer be permitted."

True enough, but not calculated to win friends and influence the people who must join together if a nationwide system of roads was ever to be built.

What Page lacked in public tact he made up for in determination and political connections. He had successfully lobbied Congress to insert $500,000 into the Post Office Appropriation Bill of 1912 for the building of more object-lesson roads around the country. At this point, the federal government still believed its role was to encourage local governments to build roads and to encourage the nation's chief industry, farming. Rural mail delivery, as a government service, didn't exist for most farmers until the late 1800s, and even then it was delivered only if the farmer lived on a good road. Bad enough that most farmers had no phones, radios, or any of the usual means of communication; but the muddy roads and lack of mail added to their isolation. The bill was the biggest step, for its time, in changing this situation. It also helped to pull the reluctant farmer into the Good Roads campaign.

The postal bill was the first time since the construction of the National Road 100 years earlier that the federal government had been directly involved in funding highways. The $500,000

**The aristocratic** Logan Waller Page antagonized many with his elitism, but he convinced Congress to take an active role in road building.

**One of the first** motorized postal vehicles. The federal government required that a county's roads be good enough to allow a postman to carry out his duties. This simple requirement became a strong impetus to improve the nation's highways.

inserted into the 1912 bill began the critical trend of federal aid for the building of highways.

The postal demonstration roads, Page convinced Congress, would act as seeds that, once planted, would grow into good roads, encouraging the counties to build more and more new roads on their own. The government predicted that with each $300 to $500 postal demonstration road it built, $2000 to $10,000 of additional highways would be built by enthusiastic state and local participants.

The program did work to some extent. Its most important outcome, however, was the lesson Page learned from dealing with over 3000 counties. After the 1912 bill, he began encouraging a partnership between the states and the federal government—bypassing the thousands of counties—and building the relationship that is still thriving today.

## THE PARTNERSHIP BEGINS

It started at 10:00 a.m. on December 12, 1914, at the Raleigh Hotel on Pennsylvania Avenue in Washington, D.C. After declaring their organization a reality, the men in the room packed up and went down Pennsylvania Avenue to the White House. There, Logan Page and President Woodrow Wilson were waiting for them.

The roads movement could have no President more favorable to its cause than Wilson. He was an enthusiastic motorist, who loved to go for drives in the White House's Pierce-Arrow. In fact, Wilson had chosen the automobile as the place to propose to his second wife, Edith Galt. Now he gave the infant group his presidential seal of approval and the American Association of State Highway Officials (AASHO) became a reality.

It may be difficult for most people to get very excited about the banding together of a group of highway officials. But in fact, the formation of AASHO

**Paving a road** was so labor intensive and required such an enormous amount of stone or brick that few roads outside of cities were paved. By the 1920s, improvements in the process of applying asphalt and cement made it easier and cheaper to pave rural sections of America.

In 1893, Massachusetts created the first statewide highway department. It was a three-man commission, charged with linking the state's major cities with one another. New Jersey was the second state to form a highway department and Connecticut was the third.

was a monumental step for the highway movement in America. For one thing, it was a professional organization made up of men—and they were all men at this point—who knew how to build good roads, were dedicated to the job and had the official authority at the state level to do so. For another, it provided a quasi-independent forum in which the states and the federal government could seriously discuss their road-building interests out of sight of politicians.

Logan Page had endorsed the move to create the organization, an idea put forward by Virginia's commissioner of highways, George Coleman. Page had agreed, with one stipulation: Membership would be restricted to the commissioners of state highway departments and their immediate staffers. Every state highway agency would have to be run by a professional engineer, "thus enabling full and frank consideration of questions, particularly those of a technical character, untrammeled by commercialism or popular prejudices." In other words, engineers and scientists only, no laymen or ignoramuses allowed.

AASHO eventually became an elite group of executive highway engineers, mostly insulated from politics and empowered by huge sums of federal funding. It would become a key player in the construction of the Interstate System.

## CONGRESS DECIDES TO PAY

The Federal Aid Road Act of 1916 was the highwatermark of Page's career. It laid the financial and political groundwork for the federal and state partnership by authorizing $75,000,000 in matching funds over five years for highway construction, the most money ever approved for federal roadwork at that time.

There was a string attached. Without an approved administrative body to interface with the federal government—Page's office—no funds would be released to any state. This was no small matter, because in 1916, 11 of the 48 states still did not have a highway department that would comply with the requirements of the act. Massachusetts had created the nation's first state highway department in 1893, but other states still needed to pass that legislation. Others couldn't come up with the required matching state funds, and still others states didn't have the right supervision of their road building, if any at all.

The federal legislation, with its promise of generous funding, forced the states to get their highway acts together. By 1919, every state had complied with the rules and formed a highway department. Now the states were on their way to becoming capable of overseeing construction contracts and interfacing with Washington.

The rules of engagement were simple, much simpler than later highway bills; but then these were simpler times. The states would decide which roads would be built and improved. The federal government would then reimburse the states for a job well done. Each state had to have an approved highway department before funding was made available. The U.S. government would pay 50 percent of the costs, and the states would pay the other 50 percent. The states would maintain the roads in perpetuity.

California was the first to complete a highway project under the new federal aid partnership. It built a two-and-a-half-mile road, 20 feet wide and paved with concrete five inches thick, in Contra Costa County.

Logan Page would not live to see the success of his efforts. He became suddenly ill while attending an AASHO meeting in Chicago. He retired to his room and died hours later. Marking his determination to the end to deliver what he felt Americans needed most, good roads, his gravestone in Richmond, Virginia, reads: "Logan Waller Page, 1870–1918, Pioneer in the Science of Road Building in the United States. Benefactor of All Who Use American Highways."

## ENTER THE CHIEF

Thomas Harris MacDonald is not what you would call a household name. Most scholars have relegated him to a footnote in American history, and MacDonald would have preferred it that way. Yet for 34 years he was the king of the road, commanding the respect of Presidents, Congressmen, and heads of state as the most knowledgeable individual in the world when it came to the building of highways.

In 1919, MacDonald was named the Chief of the Bureau of Public Roads (BPR), succeeding Logan Page. For the next three decades of his federal service, he would be called the Chief.

*The federal legislation, with its promise of generous funding, forced the states to get their highway acts together. By 1919, every state had complied with the rules and formed a highway department.*

*Above:*
**The legendary** Thomas MacDonald, a master road builder, but a man so formal he refused to shed his impeccably knotted tie even in Indian headdress.

*Below:*
**MacDonald in a pose** more comfortable for him.

MacDonald was a man formal and serious to the point of eccentricity. He had few close friends. Employees and colleagues called him Mr. MacDonald or Chief, never Tom. If the elevator door opened and the Chief was in its cab, you were forbidden to step inside, unless you were a member of his senior staff or Miss Fuller, his long-time secretary and eventually his wife.

In 1919, when MacDonald took office, a well-informed driver may have been able to find an entirely paved route between New York and Washington, D.C. However, a trip between Washington, D.C., and Richmond, Virginia, had only 100 miles of pavement, and that was only if the driver was willing to go out of his way to find it. The shorter route was almost entirely dirt. The only continuously paved roads in the rural United States were in the Northeast and along the Pacific Coast, but most of these roadways were 14-foot-wide, single-lane surfaces that forced approaching vehicles to swerve to the shoulders in order to pass safely. Poorly built roads with mottled, bumpy surfaces kept automobile speeds to just 25 miles per hour, and trucks crept along at less than 20 miles per hour. Life on the road was painfully slow, and the isolation of a large part of the citizenry from the rest was still a dark reality.

MacDonald was a son of rural Iowa, and he understood the evils of bad roads. And as Iowa's former chief engineer, he also understood the mechanics and politics of road building.

When MacDonald took over the Bureau of Public Roads, it was a well-funded but underperforming agency. The tens of millions of federal dollars set aside for road construction had been largely unspent. Disruptions due to World War I were the primary reasons, but Logan Page's own personality had also impeded progress. His insistence that road building was a job only for scientists and other professionals ruled out many knowledgeable people. Moreover, Logan believed unpaved roads were good enough for rural folk. He believed that mass transportation would always be provided by the railroads. Roadways were meant mainly to get people and produce to and from the railroad stations, and for getting the mail to the farmers. In the end, Page delivered only eight miles of new paved road to the American public.

Assuming office, the Chief quickly turned things around. MacDonald was inclusive, not exclusive. He helped win the confidence of the construction community by assembling a powerful coalition of contractors, highway officials, associations, and manufacturers. He streamlined building methods, further professionalized management on the state levels, and introduced data-gathering techniques so that roads could be constructed where demands were strongest. He believed that highways, not railroads, were the nation's transportation

future; and he systematically tracked the demise of the railroad industry, logging in the BPR records each time a line was decommissioned.

Acting more like the Commander in Chief than the Chief of the Bureau of Public Roads, MacDonald declared war on bad roads. "Those who do not have qualities of manliness, square dealing, good temper, and ability to get along with people must go," he decreed. MacDonald's call to battle was "This is an All-American Job!"

The Chief launched into an unprecedented road-construction campaign, releasing enormous sums of federal monies for the work. His new staff, somehow, approved 90 percent of state highway plans for funding within four days of their arrival at the BPR offices.

The Chief liberated nearly a quarter of a billion dollars of decommissioned World War I hardware for the construction of his highways. Steam shovels, hand shovels, and just about anything else the U.S. Army was willing to part with went to build roads, a vast improvement over the mules and wagons that had been used up to that time.

He helped secure over 40,000,000 pounds of explosives to clear the way for his new roads. The states took to this $10,000,000 arsenal with great enthusiasm. As one highway engineer explained, "The results of TNT in rock blasting are so

**Indian footpaths** became horse-and-buggy trails, which became America's first roads for cars. As recently as 100 years ago, there wasn't a single numbered highway between states, nor standard signs or maps to guide travelers.

far superior to those of any other explosive that we have found that an experienced powderman who has used TNT can hardly be induced to use anything else."

MacDonald never missed an opportunity. In 1921 the U.S. Army accepted his invitation to determine the most important overland routes for national defense. MacDonald had the Geological Survey create a set of maps for the War Plans Division to study. Large base maps were drawn up at a scale of one inch for every eight miles. When the allocated funds for the effort ran dry, MacDonald put five cartographers on his own payroll and finished the job.

In the end, a set of drawings adding up to a 32-foot-long map of the United States was delivered to the army General Staff. General Pershing himself, America's World War I hero, testified before Congress that same year that the highways most important to National Defense were those most needed for commercial and industrial growth and the motoring public. Translated—the very highways the Chief was seeking support in building. The military's map became known as the Pershing Map, and it foreshadowed the U.S. Interstate System that would develop later.

In 1921, a scant two years after MacDonald took office, he negotiated the landmark Federal Highway Aid Act of 1921. The act solidified the

**By the 1920s,** farmers were turning into merchants and innkeepers. Americans driving along roads rarely traveled were creating a demand for gas, food, and lodgings, and poor farmers were eager to accommodate them.

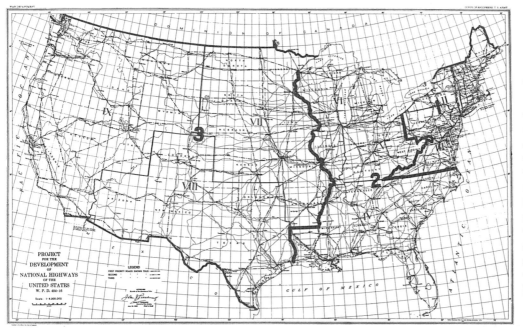

**The 1922 Pershing** Map reinforced the connection between roads for commerce and for national defense. Working with the Bureau of Public Roads, the War Plans Division identified the roads most important to potential military operations. These roads eventually became the first interstate highways.

state and federal partnership by allowing the states to manage their own road building with a budget of over $75,000,000 a year, five times what the 1916 act provided. Even the extremes of road-building factions, industry and agriculture, were pleased. The Department of Agriculture, from which MacDonald's BPR was still run, was able to promise farmers better roads, as half of the money would be spent on rural farm-to-market routes. The AAA and other promoters of grand transcontinental highways were happy too, as the other half of the money was to be spent on roads considered interstate in nature.

And that was the far-reaching provision. For the first time, the states would be committed to identifying, and building, interstates.

As the 1920s roared on, the Chief and his tribe entered the golden age of road building. In 1922, MacDonald and his partners completed work on 10,247 miles of road, more than three times the work of the previous five years. Red tape in the approval process was nearly eliminated, standards in design and construction were raised, and the BPR became the place to seek leadership and advice from experts. When a highway-safety committee discovered that an alarming number of fatal accidents occurred at railroad crossings, the Chief and his men at the BPR led the effort to eliminate the worst of these hazardous inter-sections and make others safer with the installation of simple warning signals. Many more lives were saved by making the roads flatter, straighter, and free of dangerous intersections.

Regardless of his office's growing significance, MacDonald's objective remained simple. In 1924, he explained, "My aim is this: We will be able to drive out of any county seat in the United States at thirty-five miles an hour and drive into any other county seat—and never crack a spring."

The Revenue Act of 1932 called for the first federal tax on gasoline. Damage to motorists' wallets was one cent a gallon. The tax, earmarked for deficit reduction, was to last only a year. The rest is history.

## CHAPTER FIVE

# A Radical Idea: Numbered Highways

The Lincoln Highway: America's first major interstate road built for cars and trucks.

**T**HOMAS MACDONALD, the "Chief" of the Public Roads Bureau, was not alone in his wish for a smooth ride along the nation's highways. By the time he took office, Americans were getting off their front porches and hitting the road at a remarkable rate. In 1905 there had been a mere 48,000 motor vehicles sold in America. By 1916, that number was 2,500,000, an increase of 5000 percent.

The government had been sluggish in meeting the demand the new automobiles created. But in America, the land of free enterprise, such voids don't last long. With the federal government lacking a strong policy for funding construction of the highways, the private sector saw an opportunity.

## THE TRAIL ASSOCIATIONS

No group had rushed faster to fill that void than the privately run Trail Associations. These associations, loosely bound groups of individuals interested in motor travel, usually adopted existing roads and gave them the names of famous patriots or national landmarks. The Arrowhead Trail wound between Salt Lake City and Los Angeles; the Dixie Highway ran between Mackinac, Michigan, and Miami, Florida; and so on.

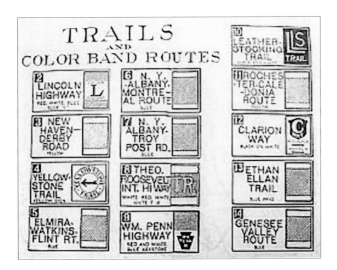

The stated purpose of the associations was to encourage motoring along their routes and, incidentally, to increase the flow of tourist dollars to the businesses along their way. The trail associations issued guidebooks, put up directions and, supposedly, helped to keep their roads in good repair. To fund their operations, they collected dues not only from motorists but also from the small restaurants, rooming houses, and tourist attractions along the road. For the larger trails, big automotive businesses like tire companies and automobile manufacturers contributed to the pot.

In actuality, most of the monies collected usually went to line the pockets of the trails' promoters. With the exception of the most successful of the trails, the Lincoln Highway, the promoter usually collected his money and did as little as he could to earn it. Some named trails were maintained, but most were not. The routes of the trails often had nothing to do with driving efficiency or even good scenery. They followed the path of most financial gain, finding their way into the towns where supporters waited for the motorists' business.

Even when the Trail Associations were operated with integrity, they were still a problem. They were competitive and uncooperative with each other. They could be fierce lobbyists, pressuring state highway departments to improve their trail over another. The routes they selected and marked overlapped each other with maddening confusion. For example, the Lee Highway and the Apache, Old Spanish, Atlantic Pacific, and Evergreen Trails all used the same path in New Mexico. One road in Missouri was known variously as Glacier Trail, National Old Trails Highway, Golden Belt Route, Daniel Boone Trail, Salt River Road, and Airline Highway.

The directional or cautionary signs along these routes were another problem. Any barn siding, rock, or tree was a fair place to put up a sign. One tree might have the signs of four competing trail associations tacked onto it. And there was, of course, no uniformity among these signs. They might be any color, size, or shape, made of any material. Sometimes the route they marked was clear and sometimes not. Sometimes they warned of hazards, and sometimes they didn't. It all depended on the goodwill and skill of the individual association.

Maps showing the relationship of one trail to another were virtually nonexistent. The American Automobile Association, founded in 1902, was one of the first to map the roads with any accuracy, struggling to pin down the wandering trails for the adventurous.

By 1925, chaos reigned on the open road. There were over 50 interstate

The term freeway usually means the highway is a limited-access roadway free of such obstructions as traffic lights and railroad crossings. The term may also refer to toll-free highway, but some freeways do charge motorists, making some freeways toll-ways!

**AAA's pathfinders** sometimes ran into trouble as they set out to find routes on which Americans could indulge their newfound love— driving automobiles.

trails and about 400 named highways spreading across the country with as much order as the cracks in a broken window. There was no uniformity in the construction or maintenance of the trails; they wandered hither and thither all over the countryside, and anyone out for a day's pleasure driving was as likely as not to get lost or, worse, crash through a fragile wooden bridge or run off a fraying roadway.

## ASSIGNING NUMBERS

In 1924, AASHO asked the Secretary of Agriculture, the Cabinet member still overseeing the Bureau of Public Roads, to find a solution to the problems caused by the named trails and their jumbled paths. The Secretary appointed three members from the Bureau of Public Roads and 21 state highway officials—all members of AASHO—to a Joint Board on Interstate Highways. The attitude of the man in charge of building the Panama Canal, General George Washington Goethals, had said that he found boards to be long, narrow, and wooden. However, this board would have impressed even the general.

The Joint Board attacked its task with a vengeance. Their job was to "cooperate in formulating and promulgating a system of numbering and marking highways of interstate character." Just another way of saying, Designate a system of interstate roads and determine how the routes should be marked. At the same time, the board had to reach a consensus among 48 independent state governments. The Joint Board's chairman: Thomas MacDonald, of course.

# Big Road Advocate

**O**N MARCH 4, 1902, nine of the country's 50 driving clubs decided to band together to fight for the right of motorists to drive over something other than muddy and rutted roads. The meeting was held in Chicago, where the local club hosted groups from Michigan, Rhode Island, Pennsylvania, New Jersey, and New York. This core of passionate drivers decided to pool their resources, calling themselves the American Automobile Association. Their combined membership was a healthy 1500, not bad for the times.

In 1902 there were just 23,000 cars moving about the United States, but those cars would not find a single paved road between any two cities in the nation. AAA began lobbying the United States Congress for transcontinental superhighways long before the general population even knew what a superhighway was. It took 14 years and a lot of convincing, but eventually AAA helped persuade the federal government to open its purse and start paying for better roads. It has never stopped pushing for the "Big Roads." And as Americans became one with their cars, AAA became one with them. Today, it is among the largest associations in the world, with nearly 45,000,000 members.

In 1915, AAA's roadside service started out as a mix between the Hell's Angels and the Guardian Angels. If a driver managed to convince friends or families to take a Sunday drive with him, there was always the question of "Will we get back, and when?" Spare tires, tubes, patch kit, air pump, gas, water, and a compass were standard equipment on automobiles of the day. But few motorists understood the complexities of the combustion engine. Recognizing this, St. Louis's AAA club decided to send out road mechanics on motorcycles to aid broken-down and stranded Sunday drivers. The men on motorcycles were called the First Aid Corps and were so popular the chapter's president said, "The club is seriously thinking of sending the first aid men out on Saturdays too."

By the 1930s, AAA had its own service division, which would make the club even more popular. Immediately after World War II, the association pioneered the use of radio-dispatched tow trucks. They were among the first to use two-way radios, thanks to a special license from the Federal Comm-

**In 1915, AAA introduced the First Aid Corps,** five mechanics on motorcycles who roamed the streets of St. Louis on Sundays, looking for drivers in distress.

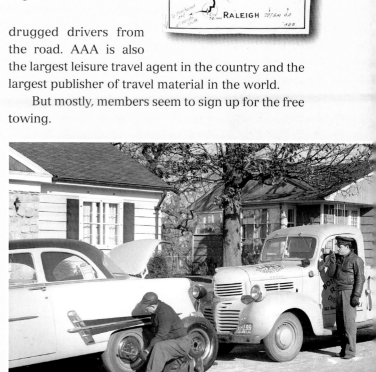

unications Commission. By 1974, AAA was defining the art of roadside service when it published the first-ever manual for towing imported cars.

## TICKING A TRIP

AAA put out some of the very first road maps for motorists. Its first map, hand drawn in ink on linen and published in 1905, was of the streets of Staten Island, New York. Six years later it mapped the entire state of New York. In 1932 the first TripTik was put out—a popular map in a booklet that allowed the driver to flip through and tick off the sections of his trip as he progressed. It was a success from the start and remained relatively unchanged for years.

Leading the world in mapping roads, AAA turned its services over to the United States Army in 1942, literally pointing the way for Ike and the Allies to drive into Berlin. By 1950, AAA owned the largest highway map ever made, measuring 67 by 100 feet and costing $20,000 to draft. These days, ticks have turned to clicks, as AAA members hit www.aaa.com nearly 500,000 times a year, often to download TripTiks.

AAA was and still is the lobbying, banking, trip-planning, insuring, and roadside guardian of the American motorist. It shamelessly pursues its members' agendas, fighting legislation that might raise the cost of operating a car. It battles to keep drivers out of harm's way by fighting to ban ever bigger trucks and fatter buses from the Interstate System. It is a vigilant advocate for eliminating drunk and

**In 1911, AAA issued** its first strip map, a forerunner to the famous TripTik. The first AAA staff and members were vigilant explorers of trails made for horse and wagon, not man and car.

drugged drivers from the road. AAA is also the largest leisure travel agent in the country and the largest publisher of travel material in the world.

But mostly, members seem to sign up for the free towing.

**In 1945, AAA was the first company** to use two-way radios to dispatch tow trucks.

**Building a better road** has long been an experiment in method and materials. This wooden-plank road, built in the 1920s in a California desert, was an early attempt to improve roads in difficult terrain. Rotting wood was the downfall of this school of thought.

Oregon was the first state to levy a tax on gasoline, in 1919. News of the tax's success in raising road building funds spread like wildfire among the other cash-strapped states. Within 10 years almost every state had a gas tax.

Some of the preliminary work had already been done. The 1921 act requested that the states select three percent of their primary systems of roads as being "interstate in character," making road connections in agreement with other states. The states had listened well and identified a total system of 81,096 miles, or 2.8 percent, for a primary national system. Meanwhile, back in Washington, the Chief and his men had conducted their own survey, systematically reviewing the country and its needs, and had come up with their own vision of an interstate system.

Now they had to get the various versions together—from all the individual states and the federal government. In a series of meetings during the spring and summer of 1925, the Joint Board hashed it out. MacDonald and the Bureau of Public Roads were there, negotiating between bordering states, resolving conflicts, and vetting out decisions based on local politics.

The states were encouraged to participate and voice concerns. In the end, attendance was nearly perfect. Every state but two was represented by a delegate or by a ballot in the form of a blank map of their state where they were to draw their desired routes.

The men who rolled up their sleeves and went to work in those meetings were not creating new roads. They were cobbling together a system from the thousands of roads that already existed all over the country, plotting routes that would allow a motorist—or a military convoy—to at last look at a map and plan a journey across the country, moving from one state to another. Many of the roads might be in bad shape, but with federal coffers open, that would change. All would be relatively narrow, with stop signs and railroad crossings slowing things down. They were a long way from being superhighways. But they were a start.

It was a formidable task, when one considers the vastness of the American continent—3000 miles through plains, deserts, and mountains, through congested cities and long stretches of land still almost a wilderness in character. Add to that the inevitable juggling of individual states for an advantage, and the miraculous nature of the final meeting of the minds becomes clear.

## DO NOT ENTER

As that meeting of the minds was emerging, another group was getting together for an equally sticky job, working out a numbering scheme—colorful names having proved too unwieldy—for the new interstate system. The five chief engineers from Illinois, Missouri, Oklahoma, Oregon, and South Carolina were the lucky men assigned to the numbering subcommittee of the Joint Board. They agreed to assemble at the Jefferson Hotel in St. Louis, Missouri, to work out a final plan. Remarkably, assigning route numbers to the nation's most important highways came down to five men in a hotel.

The word was out that an interstate highway system was in the works, and every hamlet in the country seemed to be angling for a numbered highway to run along its main street. The Trail Associations wanted their say too, lobbying to somehow keep their trails in existence. But they were locked out. Some complained about the legality of the Joint Board's authority to number the highways, but to no avail. The door to that hotel room remained firmly closed.

The board had a very good reason for wanting to operate without the interference of the named-trails' advocates. They wiped the trails out. Despite lobbying, the board refused to assign major trails, like the National Road and the Lincoln Highway, a single number. For example, some Trails Association members hoped the Lincoln Highway would be relabeled as U.S. Route 30 along its entire way. Instead, the major trails were broken up and allocated out among multiple U.S. routes, ensuring their demise.

## THE FINAL PLAN

The United States is roughly twice as wide as it is from north to south. The map the group of five used to determine the numbering system showed highways crisscrossing the country as they made their way north and south, east and west. The natural grids created made for easy identification of the routes. They agreed on the following rules.

**North–South U.S. Routes** would carry odd numbers only. The lowest odd number was the route on the East Coast, with the numbers increasing until the highest odd-numbered route was reached on the West Coast. For example: U.S. 1, running along the Atlantic seaboard between Maine and Florida, was the lowest primary odd-numbered route; the highest primary odd-numbered route would be U.S. 101, running along the Pacific seaboard between Washington and California. The primary north–south routes were to end in a one. Because there were too many major north–south highways in the plan, fives were used to signify other key north–south thoroughfares.

**East–West U.S. Routes** would end in even numbers only. The lowest primary even-numbered route was U.S. 2 on the northern border, with the numbers increasing until the highest primary even-numbered route was reached along the southern border. Transcontinental and other primary east–west U.S. routes were to end in a zero. The highest principal even-numbered route would be U.S. 90, running along the Gulf Coast and the Mexican border between Florida and Texas.

Less important through routes were to have two-digit numbers, ending in even numbers for east–west traffic and in odd numbers for north–south traffic. Spurs from the main highway would use three-digit numbers.

**Heralding**
standardization: At last! A set of standardized signs drivers could rely on from town to town and even state to state.

72

Considering the potential for growth, the gang of five intentionally left some numbers unused. Interstate Routes 8, 33, 66, and 88 were some of the numbers not assigned to the system, but they could be used later for expansion.

The numbering system was conceived with an imperfect world in mind. There were exceptions to every rule. Mountain routes, for example, because of their restrictive geographic conditions, might carry two route numbers along one highway.

The most prestigious routes for passing through a state were the primary transcontinental routes that ended in zero or one. The only serious threat to the numbering scheme came when Kentucky argued bitterly for a U.S. route ending in zero. The state claimed discrimination when it was left without a major east–west route when all the states to their north and south had one or more.

## SIGN FLAWS

While the group of five in St. Louis was assigning numbers, a second subcommittee was busy devising a uniform system of signs for the new interstate system. The confusing array of signage on the country's roads was not only inconvenient; it was dangerous.

By the mid-1920s, approximately 20,000 people a year were being killed on the highways. The injury rate will never be known for certain, as only eight states required that accidents involving injuries be reported, and about half the states kept no records at all on traffic safety.

The states of the Mississippi Valley, taking matters into their own hands, had devised a system of signs. In the absence of a national signage agenda, officials from Indiana, Minnesota, and Wisconsin traveled about the Mississippi River Valley, proposing to the various states the use of signs that were consistent in size, shape, and roadside location. Once in agreement, they posted signs throughout the region. Round signs appeared at railroad crossings, octagonal stop signs were positioned at intersections, and diamond-shape signs carried warnings on dangerous curves. The group reported its success to AASHO.

Leave it to a group of chief engineers to create homogeny. Obviously influenced by the Mississippi Valley group, the signage subcommittee came out with designs that were simple and to the point. Their handiwork can be seen along roadsides today in the now familiar black-and-white federal shield marking all U.S. routes. Borrowing from the Mississippi Valley states, the subcommittee on signage determined that danger or stop signs should be octagonal, railroad warning signs should be round and two feet in diameter, caution or go slow signs should be diamond shape, and informational signs to inform drivers about schools, hospitals, and other public facilities should be two feet square. These signs were to have a lemon-yellow background with black lettering and one-inch black borders. Directional and speed limit signs were to be rectangular with white backgrounds and black numbers, letters, and borders.

Meanwhile, when the Joint Board finished its recommendations, it had drawn up a list of 145 U.S. routes

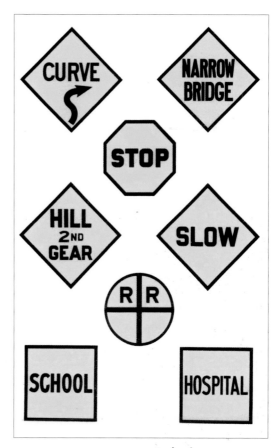

traversing 75,884 miles of highway, using a set of standard signs. Surprisingly, only five proposed paths of roadway were disputed, requiring additional negotiations between the states.

But one of those disputes endangered the whole system. It was that troublesome question in Kentucky. And it was all over two little numbers . . . or four, really. Was Kentucky's interstate going to be known as Route 62 or Route 60? The resolution of this problem would have an outcome no one could have predicted, and it resulted in the birth of one of the most celebrated strips of roadway in the world.

The committee that had met in St. Louis to decide the numbering of the interstates designated the road going from Ashland, Kentucky, to Ozark, Missouri, as U.S. Route 62. Kentucky strenuously objected to the plan. The state wanted one of the big transcontinental highways ending with a zero to run from Newport News, Virginia, across its borders and on to the west, thus ensuring a healthy amount of interstate traffic and business. They wanted nothing to do with a U.S. Route 62 that dead-ended in the Ozarks.

It was obvious, Kentucky said, that it had been blatantly bypassed. There were six new main east–west interstates to the north and three to the south,

**Early signage** was inconsistent in placement, color, size, and print, making travel a frustrating and even deadly experience. These simple standardized signs would save time and lives by informing drivers of what was ahead.

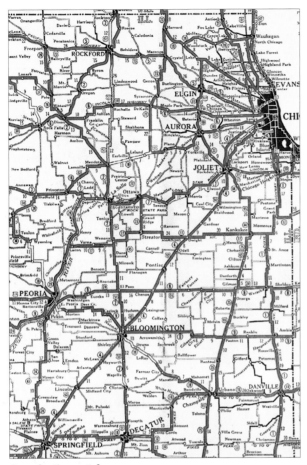

**America's most famous** highway, Route 66, was almost named route 60. Called the Great Diagonal Highway, Route 66 ran between Chicago and Los Angeles, covering 2297 miles.

every one of them ending in zero. The committee of five that planned the paths of these important highways took the U.S. route number that should have gone through Kentucky—U.S. Route 60—and hijacked it to Chicago. (As it turned out, three of the five committee members who made this suspicious decision were from states that the new U.S. Route 60 was to pass through—Illinois, Missouri, and Oklahoma.)

Kentucky became the lone dissenter to the federal plan, officially rejecting the numbering of the interstate highways. The fragile agreement between the 48 states was in jeopardy. The numbered highway system was a 100 percent volunteer arrangement, entitling Kentucky or any of the other states to walk away from the alliance at anytime.

Governor William J. Fields of Kentucky and members of his congressional delegation in Washington, D.C., met with none other than the Chief. After hearing Kentucky's case, MacDonald put his weight behind its proposal. AASHO followed, supporting Kentucky in her plan for running the coveted U.S. Route 60 through the state, connecting Newport News, Virginia, with the Ozarks in Missouri.

Missouri, Oklahoma, and Illinois and the other five states that formerly held the U.S. Route 60 were not happy. They too wanted nothing to do with a U.S. Route 62. The Chief and his minions were besieged by angry states, and it was all over a few numbers.

Finally, John Page, Oklahoma's Chief Engineer, pointed out that U.S. Route 66 was not used anywhere in the country and was available for the Chicago–Los Angeles highway. The main characters in the drama agreed that a U.S. Route 66 was preferable to a U.S. Route 62. If Kentucky would agree that U.S. Route 60 would end in Springfield, Missouri, and not continue through to the West Coast, they could in turn live with U.S. Route 66. And so a legend was born.

In the end, after the submission of 132 legitimate requests for additional miles of new interstate highway, the total mileage of U.S. routes was brought up to 96,626.

The final plan for America's first interstate system was adopted by the states voluntarily through AASHO and made public on Armistice Day, November 11, 1926. The log of new interstates first appeared when AASHO issued it in its publication "America's Highways" in April of 1927. Soon, gas and oil companies were using the U.S. routes to encourage driving. Texaco's trip map of the "United States' Federal Highways, Their Junction Points and

# The Rise and Fall of a Highway Legend

**A**T FIRST U.S. ROUTE 66 was just a simple road, starting at Lake Michigan and ending at the Pacific. It was mostly dirt, with some sections covered with bricks, asphalt, and even concrete—the most advanced material of the time. Very quickly it became a busy highway. Wealthy thrill seekers, truckers, and itinerant workers began flowing down its course. Frank Lloyd Wright, the famous American architect, said, "Route 66 is a giant chute down which everything loose in this county is sliding into southern California."

Its popularity brought on heavy volumes of traffic and innovations to accommodate it. In August 1929, the first section of U.S. Route 66 to become a divided highway appeared in St. James, Missouri; and in August of 1931, the first cloverleaf interchange west of the Mississippi River opened on a piece of Route 66 in Missouri's St. Louis County.

The stock market crash of 1929, coupled with the ill winds that stirred up America's Dust Bowl, were a double whammy for America's farmers. In the early 1930s, farmers began a desperate flight from the bone-dry fields of the prairies to the promise of greener pastures in California. Migrant workers in Oklahoma and Kansas packed up whatever they could carry and drifted westward along the path of Route 66.

John Steinbeck's *The Grapes of Wrath* gave the road one of its more popular names:

The Mother Road. Steinbeck wrote, "66 is the path of a people in flight, refugees from dust and shrinking land, from the thunder of tractors and shrinking ownership, from the desert's slow northward invasion, from the twisting winds that howl up out of Texas, from the floods that bring no richness to the land and steal what little richness is there. From all of these the people are in flight, and they come into 66 from the tributary side roads, from the wagon tracks and the rutted country roads. 66 is the mother road, the road of flight."

Just after World War II, an up-and-coming songwriter, Bobby Troup, headed west for Hollywood in his green 1941 Buick convertible, with his wife at his side. Troup was going to make it big in the land of the movies.

While driving on the Pennsylvania Turnpike, his wife asked if writing a song about the nearby U.S. Route 40 was a good idea. Troup thought otherwise; but after passing through St. Louis on U.S. Route 66, his persistent companion whispered to him, "Get your kicks on Route 66."

Something clicked, and Troup started putting the song together in the car, tracking the landmarks and towns along the way. He wrote, "It winds from Chicago to L.A., more than two thousand miles all the way." "Gallup, New Mexico, Flagstaff, Arizona, and of course Winona." Travelers could indeed get their kicks on Route 66.

Once in Hollywood, Troup performed the piece for Nat "King" Cole, who fell in love with it. Cole later said it was the song most requested of him. Looking back, Bobby Troup said, "Route 66 changed the course of my whole life. I will treasure my trip forever." He was not alone.

Americans were flocking to the new interstate highways, especially Route 66.

At the same time that Troup was penning his hit tune, Jack Rittenhouse was writing *A Guide Book to Highway 66*, which mile by mile described the gas stations, diners, and places to see along the route. With the Route 66 Highway Association and others promoting it as the ultimate road trip, travelers hit the road to

check things out for themselves. Waiting for them were alligator farms, caves, canyons, wilderness parks, mom-and-pop diners with homemade foods, and blue-plate specials. The term Americana took on a new meaning with the help of 66.

But in the end, its very popularity contributed to its demise. The post–World War II economic boom brought heavy traffic along Route 66. The gridlock became unbearable in its downtown sections, and its narrow two-lane rural sections became unsafe. The beloved 66 got a new, unflattering nickname: Bloody 66. The halcyon days of Buzz and Tod, who spent 116 TV episodes traveling Route 66 in their Corvette, were quickly coming to an end. Overburdened by its own success, Route 66 was eventually overtaken by the U.S. Interstate System.

## THE END OF THE LEGEND

In 1956, after 12 years of wrangling, the United States Congress passed the Federal Aid Highway Act of 1956 creating the funding for what is now called the Dwight D. Eisenhower System of Interstate and Defense Highways. Missouri was the first state to award a contract under the historic act, converting a piece of old Route 66

**A rite of passage:** The most celebrated highway, Route 66, drove its way into the nation's heart, idealizing the all-American road trip.

into a section of a new Interstate Highway. Like a candle burning at both ends, Route 66 began to shrink as sections built on new locations or rebuilt to more modern standards were folded into the new U.S. Interstate System.

By September of 1962, the U.S. 66 National Highway Association came to grips with the inevitable. U.S. Route 66 would be replaced by the new Interstates. In a last attempt to recapture the glory days, the association decided to request AASHO to collapse the names of several other Interstates into the name I-66, which was not yet taken. The mentality was, if you can't beat the Interstates, you may as well join them. But it was to no avail.

Interstate I-66 was eventually placed where it belonged numerically based on geography. Today, it's a short Interstate connecting the suburbs of northern Virginia and downtown Washington, D.C.

*Route 66* **first aired on CBS** in 1960 and ran four seasons. The show fueled the country's passion for the automobile in a time when 90 percent of Americans traveled by car.

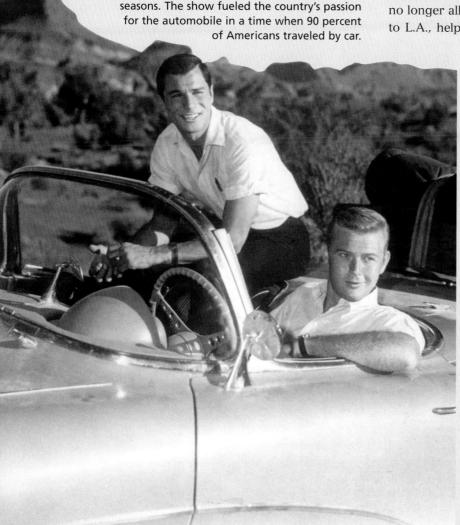

U.S. Route 66 was decommissioned piecemeal as Interstate construction progressed. California was first, wiping out U.S. 66 from Santa Monica to the California-Nevada border, eventually replacing it with I-10, I-215, I-15 and I-40. On January 17, 1977, at the other end of the highway, Route 66 was decommissioned from the Loop in downtown Chicago to Joplin, Missouri. The day was marked by a low-key event in the Windy City as the Route 66 signs were sadly removed.

In 1985, U.S. Route 66 was officially decertified when its last 1162 miles were eliminated, ending the famous highway's life at 59 years of age.

And yet, the old road somehow survives. "The death of Route 66 was the best thing that happened to it in years," says Richard Weingroff, a historian at the Federal Highway Administration. "National news coverage of the signs coming down revived interest in the route, leading to the creation of state Route 66 associations. Even the fact that its remnants no longer allow a complete trip on 66, from Chicago to L.A., helps preserve the legend. It inspires hunts for old segments, lost bridges, and intact pavement and travel guides."

In August of 1999, President Bill Clinton signed a bill into law authorizing $10,000,000 to be released over 10 years for preserving the cultural resources of the old road. As Jim Powell, a founder of the Route 66 Association of Missouri explains, "We are finding that many people are willing to get off the Interstate in two- and three-hour chunks to experience America at the slower pace of the old two-lane highway. There's something magical about the old highway that goes beyond the old, wavy lanes and the magnificent landmarks. What sets 66 apart is the people along the highway. They're always willing to help a fellow traveler. This is what Route 66 is all about . . . and why it's worth preserving as part of our heritage and history."

**The travel and** tourism industries were born when the nation's highways improved.

Mileage" showed every route laid across the continent. At about the same time, Phillips 66, a new line of gasoline, displayed a likeness of the federal U.S. route shield in its logo.

Not everyone loved the new efficiency of the numbered highways. The *New York Times* reported, "The traveler may shed tears as he drives the Lincoln Highway or dream dreams as he speeds over the Jefferson Highway, but how can he get a 'kick' out of 46, 55 or 33 or 21. The roads of America would still be on a paper if the pleas that were made ten years or more ago had been made in behalf of a numerical code." Another pundit asked, "If numbering our highways was so efficient, why not also number our rivers, mountains, cities, and Presidents?" Who would have guessed that the country would fall in love with its numbered highways.

## A PRESIDENTIAL MANIA

During the 1930s, cash—in the form of federal aid—continued to flow into the country's road-building machine even as the Great Depression fell over the land. From 1933 to 1938, an army of up to 480,000 workers was on the federal payroll at any one time, building roads. Franklin Roosevelt's New Deal spent more money on road building than ever before— one billion dollars from 1933 to 1940—providing millions of man-hours of work, desperately needed by hundreds of thousands of Americans.

In 1938, President Roosevelt called MacDonald to the White House and handed him a map of the United States. Roosevelt had hand-drawn six lines on the map, three running north and south and three running east and west. He asked the Chief to explore the viability of the federal government's building six new interstate highways as marked.

President Roosevelt was keenly interested in highway planning. Before becoming New York's governor, he had been the chairman of the Taconic State Park Commission, immersing himself in the minute details of the parkway's construction, right down to the picnic table designs. Now he was more than just curious about the feasibility of his interstate highways proposals. He saw them as a potential solution to the increasing transcontinental motor vehicle traffic and as a jobs-creation program.

The Chief returned from the President's office and handed the map to his trusted confidant Herbert Fairbank to begin working on the country's first official report on national traffic patterns.

# Roads of Ruin

**A**MERICA'S HIGHWAY SYSTEM has been a work in progress, as engineers perfected the art of road building. It has been a long and bumpy journey.

Just 100 years ago, dirt was the primary ingredient in every road in the nation. Only city streets enjoyed a top layer of bricks or stones. Farmers worked to keep the roads, their only physical connection to the outside world, in as good shape as possible. Their tractors were the first generation of road-building equipment.

**Rollers like this one**, that were used to compact a road's surface, were often powered by steam. Even today they are often referred to as steamrollers though they run on gas.

The quality of your dirt mattered the most. If the region in which you resided was blessed with high-quality ground conditions and hardy materials like stone and granular soils, you were in luck. If not, you had to make do with whatever you had, in some cases improvising with shifty clays, sand, grass, and even seashells.

**Loaders like this were** used to remove "spoil" or poor material from a roadway's path. A crude engine turned the gears and chain that moved a thick belt. Earth materials were placed on the belt and loaded into the back of a waiting horse and wagon or, in this case, a truck.

**A first-generation** tractor and grader machine, dating from the early 1900s. Driven by a primitive engine called a power unit, the early grader is surfacing a newly built road. Positioned behind the large front wheels is a row of metal spikes, called scarifiers, which gouged and loosened the road's surface so the blade following it could scrape and smooth the way.

**A pre–World War I** grader pulled by a more advanced power unit. Early graders were without engines of their own. Not until the early 1930s were grading blades and engines placed together on one piece of equipment.

**Removing haphazard** and confusing signage. With the introduction of U.S. route signs, travelers could follow a set of clearly marked numbers almost anywhere in the nation, a major breakthrough in travel. This painting, by Carl Rakiman, is part of a series on the history of roadmaking commissioned by the Bureau of Public Roads.

Even before that document was ready for release, Roosevelt sent a second map to the Chief. He had drawn two more north–south lines, increasing the total of transcontinental highways he wanted to eight. In something like a superhighway fever, President Roosevelt envisioned a network of sleek new highways, replacing the interstates then in existence. Gone would be the stop signs and traffic lights, the railroad crossings that still existed on most interstates. In their place would be wide boulevards of highways to host nonstop traffic around the country.

To pay for these glamorous new routes the federal government would create a special agency, the Federal Land Authority, to seize property needed for the highways' rights-of-way. In a process called "excess condemnation," far more land than was actually needed would be taken. Roosevelt was not thinking of the relatively narrow strips of commercial land that lined some highways. He envisioned wide swaths of acreage given over to the federal government. The extra land would be leased by investors who would build gas stations, garages, and restaurants. Their rent checks would be used to pay the government back for the cost of building the highway. The President wanted the Chief to research the idea and report back to him.

The notion of a federal agency taking over state and private properties was anathema to conservatives, who attacked the plan after a special presidential press conference on the subject. The Republican National Committee issued a statement saying it was "another ascent into New Deal jitterbug economics, but, if adopted would be the first major step toward state socialism under which the federal government would take over private industry and the United States would become a totalitarian nation."

The debate on how to fund a superhighway would rage on for almost 20 more years, but the inspiration to build one was growing strong, regardless of the rhetoric.

In the end, Roosevelt supported the Chief and Fairbank's less exciting but far more practical report, which they called "Toll Roads and Free Roads." This document, eagerly awaited on Capitol Hill, showed that most traffic in the country was in and around the cities. Vast new superhighways, running through relatively unpopulated areas, were not necessary, it said.

But Roosevelt continued to be interested in interstate highway planning right up to his death in 1945. His early maps became lineal ancestors of the Interstate System. Unfortunately that first hand-drawn map was lost.

## EXIT THE CHIEF

When Harry Truman took office, he asked MacDonald to stay on past what would have been his normal time for leaving office. But as the years passed, MacDonald found himself on the wrong side of too many arguments, especially when it came to the issue of building new superhighways in far-flung areas.

The Chief was not against superhighways, but he was convinced they were only necessary in and around the cities. His research, for which the Bureau of Public Roads was famous, showed that only a relatively small number of transcontinental trips was made each day. The traffic in rural areas simply didn't justify superhighways.

And how was the country to pay for those glossy new roads, even if it decided to build them? There were some who believed tolls could be charged. But MacDonald, a thrifty Midwesterner, did not believe that Americans would pay to drive on a road, no matter how wide and unimpeded, if a free one was available nearby.

Even the Chief underestimated the desire of Americans to get on wide highways with fast cars.

MacDonald built, paved, and improved over 3,000,000 miles of roadways, leaving Americans more connected to one another than they had ever been before. By the time he was through, almost everyone in the country lived within 10 miles of a highway. But he saw an interstate highway system dotted with traffic lights and railroad crossings in rural areas. In some cases, he was convinced that the system of U.S. routes needed only to be widened and improved, not rebuilt from scratch. He had become accustomed to seeing highway matters on his terms. It was time to make room for new ideas.

Within weeks of President Eisenhower's taking office in 1953, the Chief was asked to resign. He broke the news to Miss Fuller, his longtime—and one can only assume incredibly patient—secretary, and said in his matter-of-fact way, "I have just been fired, so we might as well get married." The Chief, not one to waste words or time, departed Washington, D.C., with his fiancée within weeks of his dismissal. In a bit of irony, they left the city by train.

Business-route markers appear on traditional interstate shields except that they are green and white instead of red, white, and blue. The markers are really just a consolation prize for towns too small to have the Interstate System pass through them but are close enough to make a short trip to their centers practical.

*81*

# The Rise of the Toll Road

**O**CTOBER 27, 1938, one could argue, marks the beginning of the modern interstates. It all started on the dairy farm of Mr. and Mrs. Eberly near Shippensburg, Pennsylvania. The Eberlys had saved the day with a last-minute agreement, selling a critical 200-foot-wide swath of right-of-way through their cow pastures, allowing the newly created Pennsylvania Turnpike Authority to start construction on the nation's first real superhighway, the Pennsylvania Turnpike.

While hosting over 300 farmers and dignitaries on their property for the historic groundbreaking, Mrs. Eberly had a last request of the officials before she turned over the property. With her five young children surrounding her, grabbing at her dress, she asked each one of the highway officials for his signature, saying, "I want these autographs so that my children can say that they saw history being made that day when the greatest highway, a new era of road building, was started." She was onto something.

**Defying the odds,** the Pennsylvania Turnpike soon carried 10,000 cars a day, far above predictions.

Twenty-three months later, Homer Romberger, a local who attended the same groundbreaking, was the first to drive on the Pennsylvania Turnpike's western end, which passed through the Eberly farm. With only 12 hours' public notice, the turnpike opened up at midnight on October 1, 1940. Up and down the new highway, locals celebrated and cheered the first car, first truck, and even first hitchhiker on the new road.

Astonishing locals, about half the drivers on the road the first day were from out of state. Motorists had driven through the night from Ohio, New York, Maryland, and West Virginia to be among the first in the country to experience the superhighway, which had been modeled after Germany's superhighway, the Autobahn . . . with no speed limits. Soon, over 10,000 vehicle trips a day were being made on this Granddaddy of the Turnpikes, doubling the projections of its critics

In their cars that first night, Americans traveled faster, farther, and more safely than ever before. The turnpike's 160 miles avoided contact with 939 local roads and 12 railroad rights-of-way along its path. It took drivers through, not around, seven mountains and over or under 307 bridges. The sweeping banked

America's Super Highway

**The Pennsylvania Turnpike's seven** original mountain tunnels were a novelty. Today, three have been abandoned. *Right:* **A firefighter** dons protective gear.

Most every state in the country wanted to build its own turnpike, but World War II didn't allow it. During the war, traffic dropped off so much on the Pennsylvania Turnpike, people were picnicking on the highway's median. After the war, however, a turnpike-building fever took hold of the nation, despite Chief MacDonald's reservations about the usefulness and expense of toll roads. Maine built the first post-war superhighway in 1947. Laid down next to the crowded U.S. Route 1, it successfully siphoned off traffic from the congested road. The success of the Maine Turnpike overwhelmed neighboring New Hampshire's 15-mile stretch of Route 1 with so much more traffic that New Hampshire decided to build its own turnpike. After only one year of construction, New Hampshire was operating the country's third turnpike. By 1954, the turnpike–toll road boom was in full swing. Many of these toll roads brought in nearly ten times what they cost to operate, proving Chief MacDonald wrong in his belief that drivers would never pay to drive on a toll road, no matter how efficient, if a free road was nearby.

curves encouraged drivers to negotiate turns at higher speeds than they had ever attempted—until this time roads were designed with flat curves to slow drivers down. Dynamite and hard work flattened out the grades of the dangerously steep Pennsylvania mountains, giving the highway graceful inclines and declines. Long 1200-foot acceleration ramps made for an easy entrance. Best of all, the entire road was designed and built in one shot. Normally road conditions fluctuated every few miles depending on the decade the surface had been built in. Now, it was smooth sailing over a consistent carpet of concrete.

General Dwight D. Eisenhower used the Nazis' own highways to defeat them in World War II.

## CHAPTER SIX

# Ike's Grand Plan

**P**RESIDENT DWIGHT D. EISENHOWER knew the value of a good road. He cursed bad ones, praised the good ones, and made a career out of understanding how to put them to use. As a young army officer he drove across the nation in 1919 as part of the U.S. Army's first motorized cross-country expedition, witnessing the sad condition of the country's roads and the gaping hole they created in the nation's defensive network. In 1922, as an executive officer in the Panama Canal Zone, Ike charted and supervised the construction of rugged jungle roads to protect the new canal from enemy attack.

As the Supreme Allied Commander in Europe, he led the largest land battle America has ever fought, the Battle of the Bulge, moving more soldiers, firepower, and supplies, at faster speeds, than the world had ever seen or has yet to see again. And he did most of it over roadways.

The successes of the road helped to make him a victorious and beloved general. That popularity propelled him to the presidency of the United States, where road building became his favorite domestic agenda.

## DIRT ROADS AND DIRT POOR

President Dwight D. Eisenhower is often misunderstood. Many have thought him to be a disengaged, laissez-faire, golf-playing politician of the rich and powerful. Perhaps his biggest failure was making his work and torment look effortless.

1902.

**Ike had a deep love** for his parents, whose lives were shaped by hardships and sacrifices. Here, Ike is on the far left.

President Eisenhower's boyhood home in Abilene, Kansas, is just two miles from the Interstate System. Once known as a dangerous place in the Wild West, Abilene is now the home of the 34th President's library, at exit 275 on I-70.

In reality, Ike grew up poor as dirt, but could recall his family's struggles with humor: "As the old expression went, the Indian on our penny would have screamed if we could possibly have held it tighter." Remaining loyal to his humble roots through his career, he was hardworking and patriotic to the core, serving not out of ego but out of a self-imposed duty to God and Country.

Ike, a name he picked up as a child, grew up in a cow town called Abilene, Kansas. Cattle were driven in from Texas and Kansas ranches to the town's rail facilities, where they were shipped east. It was a simple place with dirt roads, wooden sidewalks, and a one-man police force. The future President grew up in a small wooden clapboard house he said was smaller than his office at the Pentagon, sharing the abode with his six brothers and parents.

Roads were a recurring theme in Eisenhower's life. Looking back, he recalled, "Paving was unknown to me for a long time. Crossings of scattered stone were provided at each corner but after a heavy summer rain the streets became almost impassable because of mud. Rubber boots were standard equipment for almost everyone. In winter, snow could practically immobilize the community. I cannot recall when hard pavement was started in town but it was not earlier than 1904 or '05, I think. Even after the streets were paved, sidewalks were still made of lumber, for the most part, and the summer storms would

**Recalling his childhood** with his six brothers, Ike said, "Our pleasures were simple—they included survival—but we had plenty of fresh air, exercise, and companionship."

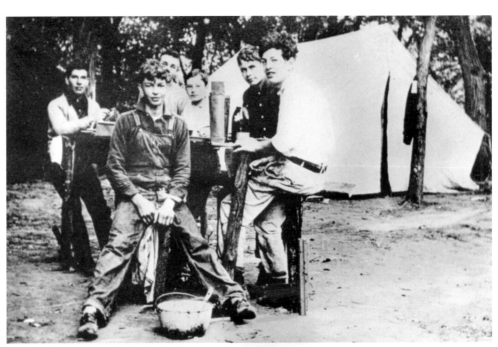

wash the sidewalks out until they were dragged back in place." It was a humble start for the world's greatest road builder.

Poor, and hungry for a college education, Ike considered the Naval Academy but settled on West Point. Before leaving Abilene, he scraped together enough money for a uniform and a train ticket and parted with his most treasured possessions: an 1897, 16-gauge Winchester shotgun and his dog Flip, which his brother Earl adopted. Arriving at West Point with only five dollars in his pocket, he found his first day at the Point was one of confusion and a little hazing.

All that was forgotten that night when he attended his first tribute to the flag as an army man. "A feeling came over me that the expression 'The United States of America' would now mean something different than it ever had before. From here on it would be the nation I would be serving, not myself. . . . Across half a century, I can look back and see a rawboned, gawky Kansas boy from the farm country, earnestly repeating the words that would make him a cadet."

West Point proved to be a tough test for the spirited, impatient, and hard-charging young man. Ike was demoted from sergeant to private for dancing too often and too closely with a young woman. He was issued demerits for rolling and smoking his own cigarettes and for being late to breakfast. He graduated with a mediocre class ranking, 61 out of 164. The one glowing review was from a superior saying he was "born to command."

Three years after graduation, Ike was drilling 10,000 soldiers in Gettysburg, Pennsylvania, preparing them for trench warfare in World War I. Before he could ship out himself, the Germans surrendered. Ike, hungry for battle, was devastated. Instead of fighting, he found himself driving across the country and into history.

**The English translation** of Eisenhower is "iron" and "hewer"—an artisan hewing metal into ornamental armor and weapons. It's an appropriate name for a West Point cadet.

## THE OLD CONVOY

Roads and trucks had saved the day in World War I, hauling men and firepower to places the trains could not. Eight months after the war ended, the U.S. Army launched its own campaign to build the highways back home that it believed were needed for national defense. To dramatize the sad state of America's roads, the army launched the first motorized military convoy to cross the country. Ike was ordered to join the convoy as an observation officer with the army's new Tank Corps.

In theory, the convoy was to make the transcontinental trip assuming an

**The leaders of the pack:**
The staff on the old convoy. The three-mile-long parade was guided by two scouts on Harley-Davidson motorcycles. Since signage on the Lincoln Highway was unpredictable and often nonexistent, the motorcyclists nailed large directional arrows to trees, fences, and poles.

*Opposite Right:*
**More than half** of the 3000-mile trip was made over dirt roads. Equipped with a tractor designed to pull cannons, the convoy was able to remove a Mack truck from a ditch.

Asiatic enemy had taken control of key railroads, bridges, and tunnels; and living conditions were to be "those experienced in the advanced zones of battle operations." In reality, the truck train became a military parade across the country. The convoy was treated like a liberating army. Over 3,250,000 Americans cheered the men and vehicles in all the 11 states and 350 cities and small towns they passed through. Greeting them were large crowds, ringing church bells, blowing factory whistles, street dancing, marching bands, and what seemed to be a politician at every stop. In Ike's words, "The speeches ran on and on."

The 3000-mile trip made its mark on the 28-year-old lieutenant colonel, shaping his understanding of roads and their role in a country's defenses. In his autobiography, *At Ease,* he dedicated a chapter to his journey on "the old convoy" labeling his memories of the sometimes hellish escapade "Through Darkest America with Truck and Tank." But he credited the convoy with laying the foundation of his belief in the need for an efficient U.S. Interstate System.

The convoy started out on July 7, 1919, when the Secretary of War, Newton Baker, ceremoniously gave the order to "Proceed by way of The Lincoln Highway to San Francisco without delay." This, however, was much easier said than done. Captain William Greany, a member of the expedition's command, admitted that the Lincoln Highway at the time "existed largely in the imagination and on paper."

Disregarding the facts and following their command, the convoy pulled out of Washington, heading north at an average speed of six miles

Motor Transport Co 554 enroute to Santa Cruz,
August 29, 1919 to escort the Pacific Fleet to San Francisco.

**Front and center:** Ike (center of picture, in light breeches) was assigned to the convoy as an observer of the Army's new Tank Corps. The convoy was an impressive show of force, with 39 officers leading 258 enlisted men driving 81 vehicles of every shape, size, and make.

*Below:*
**An ad for Goodrich** truck tires, which were used on the convoy, portrays the trip's highlights. Perhaps the most celebrated convoy in American history, it also had its hardships with a reported 230 calamities—one every 14 miles.

**The convoy damaged** or destroyed 88 bridges on its trip. This bridge survived, but the boat and its five-ton trailer were too heavy to be pulled beyond Nebraska, where it was left at the army's Fort Omaha.

per hour. They departed from the Zero Milestone that today still stands on the edge of the White House's South Lawn. It is the point from which all highway miles in the United States were to be measured. The journey was a struggle from the start, covering just 58 miles per day on average. It was the road, not the vehicles, that limited their progress.

The convoy stretched out over three miles once it was under way. An impressive show of force, with 39 officers and 258 enlisted men driving vehicles varying in size and shape—including 37 giant cargo trucks weighing up to 10 tons, 10 midsized delivery trucks, five ambulances equipped with 150 pounds of sandbags to simulate wounded soldiers, a searchlight vehicle, 11 automobiles for the officers including a Cadillac for the Expeditionary Commander (which weighed more than most of today's SUVs), nine motorcycles, four kitchen trailers, and even a five-ton trailer carrying a pontoon boat called the *Mayflower II*.

Because of the wretched condition of the roads, the most important vehicles in the convoy were two artillery tractors. Instead of pulling cannons, these two machines practically towed the convoy to San Francisco. Stationed at the rear of the procession and playing the role of sweeper, the tractors rescued heavy and light vehicles from the mud, ditches, creeks, quicksand, and deep ravines that were constantly claiming them.

More than half of the 3000-mile trip was made over dirt, mud, and sand roads. Sometimes it was impossible to find a sign showing the way to San Francisco. As a result, two "pilots" were ordered to move out ahead and blaze a trail . . . with painted arrows.

It's hard to say if the Lincoln Highway or the convoy suffered greater losses. No one was killed, but there were injuries. Trucks crashed through bridges and into rivers. They skidded off roads and rolled down mountainsides or just succumbed to the beating from the rutted roads. In the end, the convoy surrendered nine trucks to the highway, leaving them behind as unsalvageable.

The lowest point of the trip, literally and figuratively, was hit in the Great Salt Lake Desert of Utah and the Fallow Sink Region of Nevada. The Lincoln Highway through this brutal stretch was called the Desert Trail; and because it hadn't rained in over four months, the convoy's trucks, cars, and even the motorcycles sank up to five feet into the soft sands. It took an entire day for the convoy to travel just 15 miles. Lieutenant Colonel Eisenhower noted in his official report, "At best, the Lincoln Highway over this portion of the country is so poor as to warrant a thorough investigation of possible routes for building a road, before any government money should be expended on such a project." He was starting to think like a chief executive officer.

On the last day of the two-month expedition, with ship whistles blowing and flags waving, two naval destroyers escorted the convoy and the ferryboats carrying them from Oakland across the bay to San Francisco. Arriving at their destination with new uniforms issued just for the final event, the convoy rolled along crowd-filled streets to Lincoln Park, the official ending point of the

*Because of the wretched condition of the roads, the most important vehicles in the convoy were two artillery tractors. Instead of pulling cannons, these two machines practically towed the convoy to San Francisco.*

When President Eisenhower launched the Interstate System in 1956, there were only 48 states. Alaska and Hawaii did not become states until 1959. Today, Hawaii has 55 miles of Interstate System highways, leaving Alaska as the only state in America without a single mile of Interstate System highway.

**Ike and Mamie**
Eisenhower married in 1916. Sadly, their baby, Doud Dwight, died of scarlet fever. Ike said, "This was the greatest disappointment and disaster in my life."

Lincoln Highway and the end of their historic crossing. Ike simply said, "The trip had been difficult, tiring and fun."

He quickly jumped a train to Colorado to see his wife Mamie and his son, who were staying at his in-laws' residence in Denver. Their little group, Eisenhowers and in-laws, decided to drive to Mamie's parents' winter home in San Antonio, Texas. "No sooner had we left Denver," Ike remembered, "than we encountered rain. Never ceasing rains. As we got into Oklahoma, all the roads were mud and we bogged down. There were moments when I thought neither the automobile, the bus, nor the truck had any future whatever." The young couple spent a week holed up in a hotel with their in-laws, waiting for the roads to dry up. Another blow struck for good roads.

Two months later, Ike submitted his official report on the convoy saying, "The truck train was well received at all points along the route. It seemed that there was a great deal of sentiment for improving highways, and, from the standpoint of promoting this sentiment, the trip was an undoubted success." He noted that nearly every officer on the convoy filed reports recommending building public support to construct better roads. Who knew the 28-year-old lieutenant colonel would one day lead the charge?

While on the convoy, Ike had made detailed notes about the Lincoln Highway's course, conditions, designs, and construction. After that trip, he spent three years stationed at the Panama Canal, building and maintaining strategic roads and jungle paths to defend the new engineering wonder from an enemy invasion. The roads, passable only by mule, were laid down in such trying conditions that they often disappeared under mudslides and jungle brush from one season to another.

In 1925 he attended the army's Command and General Staff School, competing against 275 of the brightest military minds in the country on how to deploy armies, often over roads. This time he graduated first in his class.

General "Black Jack" Pershing was so impressed with Ike's knowledge that he selected him to create the official record of the American Armies in World War I. To do this, Ike went on a yearlong road trip studying France's battlefields and famous well-built roads. He even wrote a guidebook about them. Fifteen years later, that intimate understanding of the French highways would help him liberate Europe from the Nazis.

*Right:* **Eisenhower was** the mastermind behind D-day. A few months before the attack, Ike observes a tank exercise with Britain's General Bernard Montgomery (far right).

*Below:* **Omaha Beach,** June 6, 1944: Churchill said it was "the most difficult and complicated operation ever to take place."

## THE ROAD WARRIOR

As the Supreme Commander of the Allied Forces during World War II, Ike's instructions were clear: "You will land in Europe and, proceeding to Germany, you will destroy Hitler and all his forces."

Knowing he would have to drive into Germany to carry out his orders, Ike packed his amphibious invasion force with road-building equipment and thousands of trucks. On D-day, June 6, 1944, at his command, some of the first vehicles to hit Normandy's beaches were bulldozers and cranes. Under deadly enemy fire, the Allied Forces built exit roads off the beach and to the nearest dry roads. Following them were thousands of trucks and jeeps. The first ones literally drove off the landing crafts and into the surf, their engines having been

waterproofed in England, and began their drive toward Germany.

Before D-day, Ike's air forces had dropped over 76,000 tons of bombs on the French railheads in Paris, crippling the Germans' ability to move troops and equipment into Normandy for a counterattack. This left the Germans and the liberating Allies to battle it out on the road. For this, Ike had another plan. Landing on Utah Beach on the first day of the invasion were 1200 army truck drivers. Within months, over 23,000 of them had come ashore. This was a time of a segregated army, and nearly 80 percent of these men were African-American. Eventually they formed the Red Ball Express, named after a legendary speeding freight train, and converted 600 miles of French roads into a superhighway supply chain. The Red Ball Express moved unprecedented amounts of fuel and ammo over the makeshift highways so American tanks could keep the Germans on the run.

In the summer and fall of 1944, using the French road network he had studied and written about, Ike's forces swept the Germans back toward the motherland,

**Turning simple** country roads into super supply highways was the job of the brilliantly effective Red Ball Express.

*Opposite:*
**In pursuit** of a retreating German army. Once they landed in Europe, Ike's forces moved so quickly, they made Hitler's blitzkriegs look minor.

pushing them into an unorganized retreat. To keep the hard-charging, gas-guzzling armored divisions rolling, the Red Ball Express formed a moving supply line from the beaches in Normandy, where supplies were continuously arriving, right up to the advancing Allied tanks. The strategic use of roads and the skill of Red Ball Express drivers saved the day for the Allies.

The deployment of the Red Ball Express over the European roads was brilliant. Ike required the trucks to have three shifts of drivers, operating the trucks 24 hours a day. Between August 29 and September 15, over 6000 Red Ball Express trucks hauled over 135,000 tons of supplies. Two separate roadways, each over 300 miles long, were converted into a giant divided highway, forming a loop that ran between the beaches of Normandy and the supply dumps near Paris. A similar second Red Ball Express route was set up from Paris to supply the advancing battlefronts around Germany. To prevent traffic jams on his highways, Ike implemented makeshift pipelines to move fuels, reducing the need to put large tanker trucks on the roads. Efficiency over the road was the order of the day.

97

While a soldier in World War II, Senator Albert Gore, Sr., led a charge against a Nazi outpost. Before he could move in, his enemies disappeared. Realizing it was their highway, the Autobahn, that allowed them to get away, the future Senator, like Eisenhower, decided that the United States must have a similar network.

It seems the spirit of the United States Interstate System originated in a makeshift, war-torn France. Limited access, divided highways, rest stops, service stations, and even police enforcement were used by Ike's Red Ball Express. Rest stops were fabricated so exhausted drivers could sleep and eat. Roving teams of mechanics rescued disabled trucks, making roadside repairs or towing them to service stations. Military police kept everyone but military vehicles off the highways to keep the trucks moving.

## THE BATTLE OF THE ROAD

On December 16, 1944, Hitler made a last attempt to divide and conquer. Pushed back to his own border and in a defensive position, Hitler unexpectedly lunged into the Allies' front line. Breaking through, the Germans created a 60-mile-deep and 60-mile-wide bulge into the Allied defensive line, giving the Battle of the Bulge its name. At its onset, it could have easily been called the Battle of the Road.

As soon as the Germans began their advance, Ike identified their objective: Bastogne, a small town in Belgium with a key intersection of seven paved roads. Whoever controlled the roads would control the battle. Hitler shrewdly timed his attack to coincide with overcast skies. The Allies called the cloud coverage Hitler's Weather because it made it impossible for Ike to fly planes and drop paratroopers into action.

In a stunning move, Ike turned to the drivers of the Red Ball Express. He ordered them to drop their loads and pick up the paratroopers of the 101st Airborne, known as the Screaming Eagles, and race them to Bastogne's crossroads. With the trucks driving all night, 11,000 men, so packed in they were forced to stand the entire trip, were in position by 9:00 the next morning, arriving just before the Germans. By the end of this hard-fought and crucial battle, Ike had moved over 250,000 soldiers into action, despite the fact that the planes and trains had been immobilized.

## THE REICHSAUTOBAHN

Before the Third Reich came to power, the Germans had begun to build a superhighway, the Reichsautobahn, known more commonly as the Autobahn. Sleek and well designed, with limited access points and an unthinkable 100-miles-per-hour design speed, the Autobahn was intended to do for Germany what the Interstate System would do for the United States. It would provide jobs during its construction and economic benefits after its completion. It would unite the remote corners of the country while connecting major cities with one another. Tourism by automobile would open up new outlying areas, and a comfortable suburban life for the working class would become possible.

More propaganda than substance, the 1930s Autobahn was a broken

promise for Germany. In 1933 Hitler and his political party hijacked the fledgling national road-building program and began laying down military highways, completing only 2400 miles. Another 1500 miles were left unfinished as war needs brought construction to a halt in 1941. The incomplete highway meant that "Hitler's Road" was filled with sudden dead ends and long, partially built, disconnected sections. Even in its incompleteness, it proved to be a powerful weapon.

Hitler fought hard to keep Ike's armies off the Autobahn; but once on it, the Allies literally used the Germans' own superhighway to chase them down. Once the Allies controlled the superhighway, they were able to force an unconditional surrender in just six weeks.

Ike's men were astounded by the Autobahn's features. Some of them had never seen anything like the superhigh overpasses designed to allow German military vehicles to pass underneath. They marveled over the strength and size of the bridges, the four wide lanes, two in each direction, and the 15 feet of attractive plantings that divided them. Everyone from the five-star general down to the enlisted men was impressed by the speeds at which they were able to move over Hitler's Road.

Ike accepted the German's surrender on May 7, 1945. One of the first

**Pride before the fall:** A German military parade over a newly opened section of the Autobahn. In the end, Ike's forces moved so quickly over the Autobahn that retreating Germans sabotaged their own bridges in an attempt to slow down the advancing Allies.

**March 1945:** Defeated German prisoners of war are herded down the median of the Autobahn en route to a prison camp.

things he did as the head of occupied Germany was order an investigation of the Autobahn. Years after the U.S. Interstate System's construction began, he recalled, "After seeing the autobahns of modern Germany and knowing the asset those highways were to the Germans, I decided, as President, to put an emphasis on this kind of road building. This was one of the things I felt deeply about, and I made a personal and absolute decision to see that the nation would benefit by it. The old convoy had started me thinking about good, two-lane highways, but Germany had made me see the wisdom of broader ribbons across the land."

## I LIKE IKE

In June 1945, the most triumphant general of the largest war ever came home to the United States for his victory tour. In Washington, D.C., Senators, Congressmen, and anyone who was someone, packed into the Capitol building to hear him speak. After a heartfelt speech to a war-weary audience, the five-star general received the longest standing ovation in congressional history. In a town where people disagree for the sake of it, everyone seemed to like Ike. He easily won the 1952 presidential election, and on January 20, 1953, he was sworn in as the 34th President of the United States.

In America, the times were good but the roads were not. Even though Chief MacDonald had convinced Congress to spend millions on roadways, it was not enough. The country was at the dawn of an economic and technological boom. People were buying bigger cars with more powerful engines, about 16,000 of them a day.

But there were too few highways able to take the country's citizens where they wanted to go at the speeds they wanted to travel. Outdated prewar roads were too skinny for the swaggering chromed sedans Detroit was producing and America was buying. They rumbled impatiently as they sat stuck at traffic lights, stop signs, railroad crossings, and congested city streets. Some of the best rural highways in the country were still two-lane highways that forced drivers to squeeze their land yachts down "suicide alleys" where head-on collisions were too common. The Bureau of Public Roads' own survey showed that 76 percent of their roads were inadequate, overused, and worn out.

Before Ike's arrival in Washington, the plan was to build a 40,000-mile interstate system by improving and upgrading some of the existing roadways. The idea was to designate the most heavily traveled routes, such as U.S. Route 66 or U.S. Route 1, as improved interstate highways, converting them into "semi" limited-access highways, widening congested sections,

straightening out curves, and building overpasses to eliminate dangerous intersections.

In 1949 the 37,681 miles of highway slated to become improved interstate highways were audited and found to be in poor shape. Their makeup was remarkably inconsistent, with different types of pavement, road widths, and bridge strengths. The average age of a road surface was 12 years, and many roadways had not had a face-lift in 20 years. Thousands of miles of highway were only 20 feet wide (the narrowest section of the Interstate System today has at least 74 feet of asphalt on it). The bridges were in good repair but not adequate for the uses demanded of them. Miles of wooden bridge surfaces still existed. Nearly 90 percent of the 40,000-mile system consisted of just two-lane roads...one lane in each direction.

When Ike became President, only 6417 miles of the highways had been improved. At that rate it would take decades to finish. Moreover, since the location of the road was not changing, widening and improving the highway would require taking down homes and businesses along America's main streets. Avoiding this disruption became a goal of Ike's. He wanted to build the best interstate system in the world, with the least disruption, as quickly as possible.

At the outset of his presidency, Ike began pushing for a brand-new highway that would be built on a new right of way. Running alongside the old U.S. routes, it would be built to the highest design standards and with the best materials in the world.

Ike had a long road ahead of him. In his first full day of residency at 1600 Pennsylvania Avenue, he took Governor Dan Thornton of Colorado through his new abode and offered up a lunch of fried

*Some of the best rural highways in the country were still two-lane highways that forced drivers to squeeze their land yachts down "suicide alleys" where head-on collisions were too common.*

chicken. With the pleasantries out of the way, Governor Thornton made it clear to Ike that the states wanted the federal government to stop building roads with money collected from taxing the gas of Coloradoans, not to mention the other 47 states. The governor went a step further to say the federal government should get out of the highway-building business altogether, leaving the money and work to the individual states.

## THE GRAND PLAN

But Ike had no intention of leaving work of such overriding national importance to the states. As conservative as his politics were, he believed a new interstate highway system should be built by a centralized body, by one big government, not by 48 individual state governments. Rome, France, and Germany had built the finest highways in the world from the top down, and so would Ike. Over the next few years he struggled to win support from the public and rally Congress to his thinking.

Ike strategically chose the July 12, 1954, Governor's Conference, an annual meeting for the country's governors, to roll out his Grand Plan. Ike's sister-in-law had just passed away, and so he was prevented from attending the meeting. Unwilling to postpone his announcement, he left prepared remarks for his Vice President, Richard Nixon to read.

Ike knew this was a hostile crowd when it came to the federal government and road building, and he pulled no punches. The governors were stunned by his message.

The President wanted to build a new 50-billion-dollar highway system, aiming to finish it in ten years. Fifty billion dollars was a huge government expenditure, but with the help of tolls, Ike believed the highway could pay for itself.

Nixon took the governors back to Ike's trip in 1919 on the convoy. He pointed out, in the President's words, "Our highway net is inadequate locally and obsolete as a national system." Ticking off the penalties, he made his point. Highway casualties were "comparable to the casualties of a bloody war," 40,000 people a year were killed on the highways. Civil suits were clogging the courts and economic losses were in the billions of dollars because of highway inefficiencies, detours, and traffic jams. Most gripping were the highways' "appalling inadequacies to meet the demands of catastrophe or defense, should an atomic war come."

The room full of governors was buzzing. Nixon concluded, "I would like to read to you the last sentence from the President's notes, exactly as it appears in them, because it is an exhortation to the members of this Conference. Quote: 'I hope that you will study the matter, and recommend to me the cooperative action you think the Federal Government and the 48 states should take to meet these requirements, so that I can submit positive proposals to the next session of the Congress.'"

In 1953, when Ike took office as the 34th President, only 53 percent of the nation's 3,000,000 miles of roads were paved. Building the Interstate System required approximately 2,000,000 acres of property.

In other words, go figure out what you want and let's build it! It was too good to be true, or so it seemed.

The governors were at first shocked, suspicious, and reluctant; but the savvy politicians saw an opportunity in the Grand Plan's fifty-billion-dollar pot of gold. The speech paved the way for Ike's plan to be accepted by the states. With who was going to be in charge out of the way and what to build coming into focus, the next big question was "Who is going to pay for it?"

## DIVIDED BY HIGHWAYS

In the following year, the need for a superhighway program became even clearer when tensions escalated between America and the Russian-Chinese alliance during the Formosa Straits crisis. Senior advisors repeatedly advised Ike to use nuclear weapons on the Communists to prevent an attack on Formosa. If nuclear bombs then started dropping on the U.S., 70,000,000 urban residents—the military's estimate of the population of the nation's cities—would have to be evacuated to safer areas. The only way to do that was over superhighways.

Still, Ike's Grand Plan proceeded in fits and starts. Figuring out how to pay for the highways that everyone seemed to want to use was the challenge. Several approaches were suggested. At Eisenhower's request, his trusted aide General Lucius D. Clay had convened a blue-ribbon panel to study the question and offer a solution. The Clay Committee came up with a proposal to spend 27 billion dollars over 10 years to build a system designed to handle the nation's growing traffic needs until 1974. Financing would be provided by the sale of bonds, and bondholders would be rewarded for their investment by being paid interest. Ike himself preferred the collection of tolls to the sale of bonds, but that was a fight for another day.

**The Clay Committee** devised the first plan for Ike's Interstate System, saying it was "the top national economic and defense priority." *Left to right:* Sloan Colt, a banker; William Roberts, chairman of a large earthmoving manufacturer; General Clay, who served under Ike in World War II and was the committee's chairperson; Stephen Bechtel, owner of the largest civil engineering firm in the world; David Beck of the International Brotherhood of Teamsters.

**Nuclear jitters:**
In 1953, the year
Ike first became
President, 79 percent
of Americans believed
Russia intended to
rule the world. A
nuclear war between
the superpowers
appeared imminent,
prompting Americans
to build bomb shelters
stocked with food
and to plan for their
escape from a nuclear
strike.

Senator Albert Gore, the first-term Senator from Tennessee and chairman of the Senate's Subcommittee on Roads, had another proposal. The populist Senator didn't like the sound of those interest payments—which would siphon billions of dollars away from road building and into investors' hands. He boldly challenged the Administration, putting forth a plan to build a 41,000-mile highway system, spending 10 billion dollars until 1961, the first five years of its construction.

However, as a Senator, Gore could not write the financial portion of his proposal. Financial components to a bill must originate in the House of Representatives, not the Senate. So Gore's counterpart in the House, Representative George Fallon from Maryland, played catch-up, drafting the House's legislation and dropping in a financial mechanism.

Fallon created a bill somewhere between the Clay Committee's proposal and Gore's bill. He suggested using gas and other user taxes to pay for 90 percent of the interstates' cost and allowed eight years for its construction. Because the military purposes were too important to ignore, he named the plan the National System of Interstate and Defense Highways.

The title would eventually stick, but Fallon's 1955 federal-aid-to-highways bill was shot down by a heavy margin, 292 to 123. Strangely, it appeared that the groups that stood to gain the most from its construction were fighting it the hardest. Tire, petroleum, and other user groups helped kill Representative Fallon's bill. In an unexpected show of force, truckers alone swamped Representative Fallon and other Congressmen's offices with a record 100,000 telegrams just before the vote on the new legislation. They were protesting what they believed were taxes that forced them to pay a disproportionate share of the

*If nuclear bombs started dropping, 70,000,000 urban residents— the military's estimate of the population of the nation's cities— would have to be evacuated to safer areas. The only way to do that was over superhighways.*

Crowds amounting to nearly 100,000 people caused a three-mile-long traffic jam on Interstate 70, near Eisenhower's place of birth, during his funeral procession in 1969.

cost of the Interstate System. In an eleventh-hour turnaround, the White House tried to save the bill with Ike's show of support for taxes instead of tolls, but it was too late.

It appeared that the Clay Committee and the Fallon bill had failed. They hadn't. They created models that were debated, scorned, and ridiculed, but they allowed all parties to air their concerns and move forward, galvanizing interested parties into a coalition for the future.

## A HEART ATTACK, A STOMACHACHE, A VICTORY

1956 didn't look good for Ike. His party had lost control of the House and Senate, which made pushing legislation for the Grand Plan through Congress a long shot. And in the fall of 1955, Ike had suffered a serious heart attack, forcing rest and some introspection.

That winter, while recovering at his farm in Gettysburg, Pennsylvania, the President was deep in thought, contemplating what the country needed to prosper and protect itself. During this time he made more entries into his diary than at anytime during his presidency. His notes were filled with concerns over long-term problems that were economic and nuclear in nature. Ike was determined now more than ever to deliver the Interstate System.

**Ike was in Walter Reed** Hospital recovering from surgery when he signed the bill authorizing the Interstate. No photographs were taken of the monumental event, but Ike was reported to be "highly pleased."

And then it all came together. Ike continued to hammer home the urgent need for a modernized interstate highway system. Tire, automobile, truck, and petroleum manufacturers fell in line with construction and travel associations. The White House and Congress, Republicans and Democrats, saw the benefits of the Grand Plan and their own stake in it. The entire bunch threw their support behind a new and improved federal-aid-to-highways act, also drafted by Fallon. The new bill established a Federal Highway Trust Fund devised by Congressman Hale Boggs of Louisiana. The trust, which gave the bill real financial strength, would collect its money from taxes such as gasoline, providing the funds to pay the states back 90 percent of the cost of the Interstate System. Fallon's work became known as Title I, and Boggs's trust fund was known as Title II.

The House passed the 1956 highway legislation act by 388 to 19, an amazing 95 percent approval. The Senate, receiving the bill from the House, put Gore's language in and passed it with a voice vote.

On June 29, 1956, while recovering from an intestinal infection and operation at Walter Reed Hospital, Ike signed the bill into law and changed the course of history. Putting aside a pen for Senator Gore, Ike had a quiet dinner with his son John and his wife Mamie, resting up on his last day in the hospital. It was later understated that Ike was "highly pleased."

## START YOUR ENGINES

The Federal-Aid Highway Act of 1956 led to the construction of more than 41,000 miles of highway. Over the course of 13 years, the original act promised to reimburse—not advance—the states 90 percent of the cost to complete their portions of the system, whatever that cost ended up to be. Money collected from federal taxes on gasoline, diesel, rubber tires, heavy trucks, buses, and other items were to be put into the Highway Trust Fund, to provide for the federal government's payments. Tolls would be allowed in certain areas. Eventually the road was required to accommodate traffic growth for 20 years. It was to have a uniform design and was to be built with the highest quality materials that could be found.

"I see an America where a mighty network of highways spreads across our country," Ike proclaimed, feeling better and on the campaign trail later that year. It didn't take long. On August 2, 1956, the first contract for the building of the Interstate System was awarded by the Missouri State Highway Commission, eventually converting a section of U.S. Route 66 in Laclede County to Interstate 44. The first ground to be broken was also in Missouri, on August 13, 1956, when the state began building Interstate 70 in St. Charles County. The first section to be completed was in Kansas. On September 26, 1956, the State Highway Commission finished an eight-mile strip of the new Interstate 70, just down the road from Ike's hometown of Abilene.

Ike had been bedridden and out of public sight during the signing of his landmark legislation, but he stepped forward for a highway-building photo opportunity at the swearing-in ceremony for the first Federal Highway Administrator, John Volpe of Massachusetts. Ike held the Bible as Volpe took his oath to serve in the newly created office. As administrator, Volpe would lead the first efforts to build Ike's Grand Plan. Never before and never again would the President of the United States attend such a ceremony for a highway administrator. Ike was making a point and seeing his program launched.

These were happy days for Eisenhower. The Grand Plan was under way, the public heaped approval on him, reelecting him in 1956 with the highest popular-vote margin in American history, 57 percent. His first term saw the start of the largest highway project in the world, and his second term would see the completion of nearly a quarter of it.

In his memoirs, Ike said of the highway-building program, "More than any single action by the government . . . this one would change the face of America. . . . Its impact on the American economy—the jobs it would produce in manufacturing and construction, the rural areas it would open up—was beyond calculation."

How right Ike was.

**The Federal-Aid** Highway Act of 1956 changed modern America's landscape forever. Just 29 pages of 6-inch-by-9-inch paper, it testifies to the power of the pen.

# America Moves to the Suburbs

**W**HEN SOVIET DICTATOR NIKITA KHRUSHCHEV visited the United States in 1959, the White House press corps asked President Eisenhower what he wanted the foreign leader to see. "Levittown," replied Ike without hesitation, because it was "universally and exclusively inhabited by workmen."

And so it was. After World War II, middle-class domestic life in the United States was given a decisive push out of the cities and into the country by two pieces of federal legislation: Ike's Grand Plan for interstate highways and the National Housing Act of 1949. The Interstate System gave Americans a convenient way to commute to and from crowded cities, and the housing act offered financial incentives that made home ownership affordable for the working middle class. Suburban America was born, and boomed, with this powerful combination of government actions.

The housing legislation was partly a reflection of Americans' deep need to own the land they live on and partly an exercise in governmental self-interest. "Children and dogs are as necessary to the welfare of this country as is Wall Street and the railroads," declared the post–World War II Presidential

Conference on Family Life. Or, as the builder William Levitt put it: "No man who owns his own house and lot can be a Communist. He has too much to do."

In 1947, Levitt and Sons began building their all-American "town" on Long Island in New York. They strategically located it 25 miles east of Manhattan on a former potato field bounded by three parkways: Southern State, Northern State, and Wantagh State. In doing so, Levitt started the trend that has dominated American living ever since: a large subdivision on the city outskirts, convenient to a highway.

With 82,000 residents in 17,400 separate houses by 1951, Levittown, New York, was the largest housing development ever constructed by a single builder. Before World War II, a typical contractor built fewer than five houses per year. Levitt changed that forever, following his New York success with Levittowns in Pennsylvania and New Jersey.

And though the first Levittown was constructed before the Interstate, it was the model that builders and developers would copy for the next 50 years. Levitt and the national road system brought the suburb to all parts of the country.

## THE ASSEMBLY LINE

If Henry Ford put the masses on wheels, Levitt put them in their own homes. He put the principles of mass production to work in home building and turned his entire site into a gigantic factory. First, workers leveled the land and cleared the trees, then dropped materials every 60 feet across the site. In effect, this

set up an assembly line. He then divided the construction process into 27 clearly defined steps, and trained one crew for each job. The first crew dug the foundation, the next set the copper pipe, the next poured the concrete slab, and on through the framers and the tile setters, the red painters and the white painters, and the crews that swept the floors for the new owners. Each crew moved house to house along the line, performing its job and only its job over and over again. By 1950, Levitt and Sons was constructing one four-room house every 16 minutes.

The king of efficiency, Levitt prebuilt as many parts as possible, trucking them to the site, where they could be assembled like giant model airplanes. Lumber, pipe, and other materials also arrived by road, precut. A concrete slab with radiant heat replaced an excavated basement, and new power tools made sawing, routing, and nailing much faster. To further reduce costs and control the process, Levitt and Sons made concrete, grew lumber, and bought appliances from wholly owned subsidiaries. Levitt was able to undercut his competitors by $1000 per house and still make a better profit. The Cape Cod model sold for $7990, the Ranch for $9500.

In Levitt's 1949 model Cape, happy new homeowners would live in 700 square feet of space, including two bedrooms, a living room with fireplace, a dining alcove, and a kitchen situated so that mom could both work inside and watch the kids play outside. An unfinished attic gave space for growing families to expand.

The Levitts also adopted the auto industry's strategy of model years. Houses were updated annually, with slight style changes. Instead of tail fins and chrome, here were washers and stoves. Appliances were sold "built-in"—a phrase coined by Levitt. The '49 model included a refrigerator, a Bendix washing machine, and a white picket fence. The 1950 model included an eight-inch television. As built-ins, the appliances became part of the house and thus part of the mortgage. Mr. and Mrs. Homeowner would pay for that TV set for the next 30 years.

## HEART AND SOULLESS

For young working families living in crowded city apartments, the instant bedroom communities built by Levitt and others offered an escape that previously only the rich could afford. Young couples flocked to buy their own private patch of lawn and freedom. Sitting in their Cape Cod house, they would watch *The Honeymooners* on their eight-inch TVs and see that cramped urban life they had managed to escape. *Ozzie and Harriet* and *Father Knows Best* represent-

**The combination of the Interstate System** and developments like Levittown created the nation's first suburbs.

ed the world they had just joined. Levittown transformed the American dream into reality.

But not for everyone. Levitt's progressive construction techniques were offset by his regressive racial attitudes. The company would not sell to black families. Not one of Levittown, Long Island's 82,000 residents was African-American.

The American dream—however limited—was part of the federal agenda. In 1949, Congress revised the National Housing Act of '34 to include this objective: "the realization as soon as feasible of the goal of a decent home and a suitable living environment for every American family." By 1957 the Federal Housing Authority was financing 4,500,000 suburban homes, or 30 percent of all houses every year. And virtually all of those homes would be located near the newly expanded U.S. Interstate System.

Road builders went great guns building the first half of the Interstate System in only 10 years, from 1956 to 1966. The second half would take 40 years to complete.

CHAPTER SEVEN

# The Interstate Decade: 1956–1966

**A**T THE HEIGHT of the Interstate System's construction, Dewitt C. Greer, a proud Texas state highway engineer, summed up the feelings of those creating the dream. He said, "This Nation doesn't have superhighways because she is rich, she is rich because she had the vision to build such highways."

And what a vision it was. For this giant engineering project, America intended to build a 41,000-mile network of superhighways stretching from border to border north, south, east, and west—straight, smooth highways, with wide lanes and gentle banked curves designed for speed. Wide medians, planted with screening trees, would separate the lanes in each direction, protecting motorists from oncoming headlights. Nothing would impede the forward flow of this royal network of highways, no rude traffic lights or presumptuous side streets would interrupt its important business.

To accomplish this marvel, America would set

During the Interstate Decade, one billion dollars of Interstate System funding provided 48,000 workers with a full year's pay.

aside over 1,600,000 acres of land, an average of nearly 40 acres for each mile. Part of that land would be for the highway itself, part for the services and amenities that were expected to grow up beside it. The entire System was meant to be completed by 1972, the year funding for Eisenhower's highway legislation ran out. Once that money was gone, there was no guarantee that Congress would appropriate any more or, if they did, that the terms would be so generous—a 90 percent reimbursement to the states for their costs.

The 1972 target date was overly ambitious, as it turned out. Work on the System dragged on into the next millennium. But during the years 1956–1966, a span that can appropriately be called the Interstate Decade, more miles of the Interstate System were constructed than at any other time. More than half of the new Interstates were built during those 10 years as money poured into state highway departments.

Immediately after signing the landmark legislation in 1956, President Eisenhower moved to get his Grand Plan underway. The states' highway planners, engineers, contractors, and assorted officials began gearing up for a job that would span entire careers. Building the Interstate System was an all-consuming effort, pulling some professionals in so deeply that its work became their life's calling. The undertaking quickly became bigger than anyone associated with it. Once started, no one could claim it was theirs. Presidents, governors, and leaders of industry became participants and spectators to the biggest show on earth.

## SETTING THE STANDARD

Devising the rules for building the System was the first step. The federal government and its 48 partnering states and the District of Columbia (Hawaii and Alaska were not yet states) began setting standards for the highway's signs, designs, and materials. Choosing the Interstate System's trademark signs and shields was as urgent a need as determining the best steels, gravel, and sands for the new highway itself. Standards on the Interstate System constantly evolve, but in 1956 the states and the Bureau of Public Roads were tasked with setting them from scratch, creating a starting point for construction.

Standardization is the Interstate's hallmark and perhaps its most important characteristic. Before the System was built, a driver traveling 10 miles could experience a variety of conditions, from the bone rattling to the sublime. A washboard stretch of highway with lumpy asphalt and steep climbs might lead to a smart new superhighway with four lanes of smooth concrete, which abruptly ended at a toll plaza and dumped the driver back onto a two-lane roadway with treacherous curves and dangerous railroad crossings.

THIS IS THE FIRST PROJECT
IN THE UNITED STATES
ON WHICH ACTUAL CONSTRUCTION WAS STARTED
UNDER PROVISIONS OF THE NEW
FEDERAL AID HIGHWAY ACT OF 1956
MISSOURI STATE HIGHWAY COMMISSION
CAMERON, JOYCE & COMPANY
CONTRACTOR

**On August 13, 1956,** Missouri became the first state to begin construction on the Interstate System. The initial work converted a section of U.S. Route 40, known as the Mark Twain Highway, into I-70.

In 1955, all parties involved in standardizing the System began meeting to try to determine everything about the Interstate System's future materials and designs. One of their first decisions was to build, somewhere near Ottawa, Illinois, seven miles of test road, at a cost of $27,000,000. The testing facility was a joint private and public effort, funded by the states, the federal government, the American Automobile Association, the Automobile Manufacturers Association, and others.

By November 1958, the test track was ready. This highway to nowhere may have been the most important stretch of the Interstate System ever constructed. It was an experiment in the durability of road materials and the benefits of one design over another. The site in Ottawa was a virtual highway laboratory, where scientists used 24-ton U.S. Army missile carriers loaded with blocks of concrete to carry out their work. Their mission: to see how long it took to pulverize the experimental roads and bridges.

The test track was seven miles long, two lanes wide, curving and straight. Half of the seven-mile loop was paved with concrete and the other half with asphalt. The track was built in 836 separate sections. The sections had various subsurface materials and used different engineering concepts. Included in the

**The 1958 AASHO** road test went on for two years, deploying a long convoy of army vehicles ranging from pickup trucks to rocket launchers. Their mission: test the limits of design and materials for the new Interstate System.

loop were 16 bridges, scaled-down models of the over 50,000 similar structures that would eventually be built on the Interstate System.

The road test was not designed to come up with a single method of engineering the entire Interstate System. It was meant to help the states determine the most practical design, materials, and methods for their individual stretches of highways. Durability was important because the System would have to last longer than any highway Americans had ever built before. Each mile was designed to last until 1975. (Later it was written that Interstate System highways be engineered to last at least 20 years before needing reconstruction.)

Each mile of the Interstate System would carry dissimilar ratios of cars to trucks, so different designs for different loads would be needed. Raw materials—clays, sand, and stone native to the various states—were trucked in. If they stood up to the tests, costs would go down. Overbuilding was expensive and

wasteful. Underdesigning could cost more in the long run if the road wore out prematurely.

The Defense Department supplied large and small vehicles for testing and plenty of soldiers to drive them. Their duty was simple: Start driving and don't stop until November of 1960. They carried out their orders and wreaked the expected devastation. Entire sections of road and bridges were destroyed as the heavy army trucks rumbled over them day after day.

The section of roads left standing were the obvious winners. They had survived 1,114,000 axle loads, varying in torment from one ton to 24 tons. The information on the best performances of steel bridge beams, concrete mixes, and asphalts was compiled into AASHO manuals, which became the bibles for building the Interstate System. These were quickly revised whenever tests showed new developments or when new discoveries came to light. The manu-

als were distributed to the states as the building of the Interstate System began. The seven-mile road to nowhere proved its worth. Decades later, highway officials are still designing and building sections of the Interstate System based on its results.

The manuals also standardized the System's design, eliminating expensive and time-consuming custom designs. Plugging in already proven design specifications and materials for a superhighway's subbase, base, and surface prevented the states from reinventing the wheel each time it designed a new stretch of road. This reduced costs dramatically in every state. In Kansas, for example, once a solid set of standards for bridge construction was decided on, the forms from one concrete bridge could be used and reused all over the state, reducing construction costs considerably.

## TRUE COLORS

**Federal Highway** Administrator Bertram Tallamy (far right) wanted the Interstate System to resemble the New York Thruway, a highway he helped build.

While the test track in Ottawa, Illinois, was being constructed, a test track of another sort was already up and running in Greenbelt, Maryland, just outside of Washington, D.C. This track wasn't testing the limits of concrete, steel, and asphalts. It was seeking the best design and color combinations for the thousands of signs along the new Interstate, signs seen from the windshields of vehicles passing at 70 miles an hour, the design speed of much of the System. The test was also meant to settle a score.

In February 1957, Bertram Tallamy became the second Federal Highway Administrator, taking over the reins from John Volpe. As chairman of the New York State Thruway, Tallamy had successfully completed the nation's longest superhighway, considered one of the finest highways in the world. His objective, as the new Federal Highway Administrator, was to stir the 48 states into an Interstate System building frenzy. However, one task stood in his way: selecting a color pattern for the large exit signs along the future Interstates.

For Tallamy, the choice was obvious: blue backgrounds with white letters and numbers. He had erected signs just like these on his thruway and was getting rave reviews for them. If it was good enough for the thruway, it was good enough for the Interstate System.

Signage on the Interstate System was extremely important. The initial investment would be $200,000,000. Each sign would cost about $10,000, and some of the big signs straddling highways overhead could cost more than $50,000 dollars apiece. Traveling at normal speeds on the Interstate gave a driver only 10 seconds to read and react to a sign 1000 feet away. Obviously, the signs needed to be as easy to read as possible.

The problem was that Bertram Tallamy, the key decision maker, was los-

**A drive-by signage** beauty contest was held near Washington, D.C., to pick the best colors for the new system. The drivers passing by the sample signs chose the green-and-white signs that are now ubiquitous on Interstates.

ing his ability to discern colors. In short, the man in charge of choosing the colors for the world's largest signage program was becoming color-blind. Nonetheless, he opposed all alternative color schemes proposed to him and insisted on his blue-and-white signs.

Meanwhile, AASHO had canvassed its members and created a committee to select shapes, sizes, and colors of signs. They disagreed with the top man's taste, preferring a green-and-white color combination, and challenged Tallamy to a contest.

To resolve the issue, a yet-to-be-opened section of highway in Greenbelt, Maryland, was sectioned off, and hundreds of drivers were asked to speed past the signs, making their choice on background colors. Mock signs were erected with I-90 and U.S.-20 shields on display. The signs, showing a little humor, read, METROPOLIS, UTOPIA, EXIT 2 MILES. The three different backgrounds were Tallamy's blue, AASHO's green, and a third entry of black. All the letters and numbers were white.

After hundreds of drive-bys, the motorists were polled. Tallamy's blue was defeated handily. More than two drivers chose AASHO's green for every one that sided with Tallamy's blue. The black sign was the least popular, with only 15 percent of the vote. Tallamy's decision was made for him.

While Americans have a nation of large green-and-white signs telling us which Interstate exit to use and how far ahead our hometown is, Tallamy would be happy to know that blue-and-white signs still live happily ever after on the Interstate System, directing drivers to services and rest areas.

The first transcontinental Interstate System route, spanning the continent from the Atlantic to the Pacific Oceans, was I-80. Its 2907 miles were opened to traffic on August 22, 1986. The dedication ceremony took place just 50 miles from where the nation's first transcontinental railroad came together with its Golden Spike.

458
DEL

ROUTE
40
KANSAS

ALABAMA
23
INTER·STATE

VERMONT
15
INTERSTATE

WHITE 10"
SERIES D OR E
NUMERALS ON BLACK

BLACK 2" LETTERS
SERIES C OR D
ON WHITE

INTERSTATE
U.S.
503
N.Y.S.

MARYLAND
340

NORTH CAROLINA
2
3
INTERSTATE

THE TRIANGLE-IN-CIRCLE IS THE NATIONAL
EMBLEM OF CIVIL DEFENSE. ONE OF THE MAIN
PURPOSES OF THE INTERSTATE HIGHWAY SYSTEM
IS TO BECOME A VITAL PART OF OUR NATIONAL
DEFENSE. THEREFORE THIS SYMBOL IS THE
FOCAL POINT OF THE MARKER WHICH SHALL
CARRY THE TWO-DIGIT DESIGNATION.

THE TRIANGLE IS ALSO A TRIUNE SYMBOL OF
THE INTERSTATE SYSTEM, MAKING IT A UNIT OF
THREEFOLD PURPOSE, CONNECTING THE BORDERS
OF THE UNITED STATES, CANADA, AND THE REPUBLIC
OF MEXICO, AS A MUTUAL BOND OF FRIENDSHIP AND
DEFENSE.

THE EAGLE IS THE EMBLEM OF OUR NATION,
AND WITHIN IS THE WORD "INTERSTATE"
DESIGNATING THE NATIONAL SYSTEM.

ELMER L. ERKKILA
MINNESOTA DEPARTMENT
OF HIGHWAYS

INTERSTATE
27

## ONE IN A HUNDRED

AASHO, always thorough, had asked its member state highway departments to submit design and color choices for the shields that would identify the highways themselves. They received over 100 entries of colorful stars, shields, and some nearly abstract designs. An executive Route Numbering Committee narrowed the 100 choices down to four.

In the summer of 1957, these four were shipped to LaSalle, Illinois, where AASHO was having its annual meeting. Along a simple country road, the Interstate System's future highway shields were planted into the ground for final selection. During the meeting, highway officials from 36 different states spent days and nights driving by the signs, in cloud coverage and sunshine. In the end, it was a draw between entries from Texas and Missouri. Instead of settling for one, the best qualities of the two entries were combined into one. The design carrying a blue shield with the number of the highway was married to the top of a shield with the word Interstate across it, creating the now famous red-white-and-blue shield seen everywhere on the Interstate System.

## THE SAME BUT DIFFERENT

Fifty years after the U.S. routes were organized, it was time to devise another scheme for numbering the new system. With 58 Interstate highways to be ultimately organized, the task fell once again to AASHO's Route Numbering Committee, whose slogan could have been "Don't mess with success." The scheme for the new system of highways was the same as the one from the 1920s… but different.

Just as with the original U.S. routes, the numbers assigned to the new Interstate System's highways running east and west were given even numbers, with the principal highways ending in zeros. The highways running north and south were assigned odd numbers, with the principal routes ending in ones and fives. As before, the states worked out the numbers with AASHO, but that's where the similarities between the U.S. routes and the new system end.

The old U.S. routes began their numbering in the northern United States, with U.S. Route 2 running along the Canadian border, and ended in the south with U.S. Route 90 along the Mexican border. The main north–south U.S. routes began on the East Coast, with U.S. 1 running between Maine and Florida, with higher numbers as a traveler moved west to U.S. Route 101, running between Washington and California. For the Interstates, a mirror image of that system was devised in order to avoid duplication, and the traveler's nightmare of having U.S. 95 running alongside Interstate 95.

*Opposite:*
**What might have been:** These eight contestants were some of the 100-plus entries submitted by states in hopes of being picked as *the* shield for the Interstate System.

*Above:*
**In 1957,** AASHO selected this Interstate System shield. It was a combination of the submissions from Texas and Missouri. Today this shield is an American icon and a registered trademark.

*121*

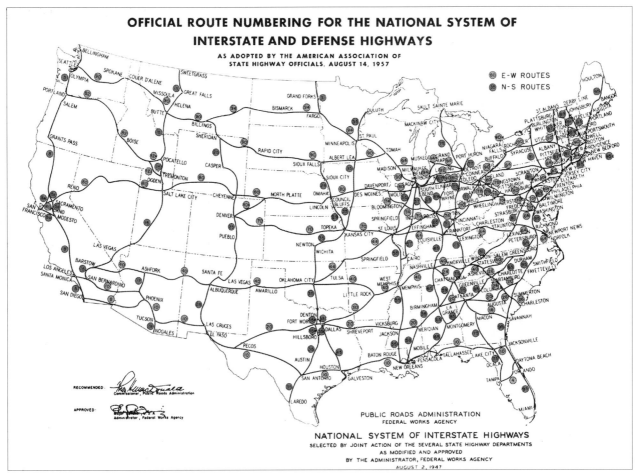

OFFICIAL ROUTE NUMBERING FOR THE NATIONAL SYSTEM OF
INTERSTATE AND DEFENSE HIGHWAYS

AS ADOPTED BY THE AMERICAN ASSOCIATION OF
STATE HIGHWAY OFFICIALS, AUGUST 14, 1957

**Avoiding chaos** in 1957. To prevent a numeric overlap with U.S. routes, the numbering of the Interstates begins in the southwestern corner of the United States. It increases the farther north and east the highways are found on the System—the exact opposite of the U.S. routes.

The Interstate System's lowest numbers begin in the western part of the country, with I-10 running west and east between Southern California and Florida and I-5 running north and south between Washington and Southern California. The numbers increase the farther north and east one goes, ending with the higher-numbered highways in the northeastern corner of the country.

No state was permitted to carry a U.S. route with the same number as a new Interstate System highway that ran through that state. This was a problem in the middle of the country, where the numbering patterns of the U.S. routes and the Interstate System were due to collide. The only way to avoid this dilemma was to rule out the use of an Interstate System highway numbered 50. This prevented having U.S. 50 and an I-50 running through California, Nevada, Utah, Colorado, Kansas, and the rest of the states clear through to Maryland. Eliminating 50 from the Interstate System has prevented countless arguments of "I thought you meant I-50 not Route 50! Which 50 are you talking about?"

Today, the U.S. routes and their black-and-white shields cohabitate well with the Interstate System's red-white-and-blue shields. They are often placed on the same signposts. This is to benefit drivers who travel over sections of highways shared by a U.S.-numbered route and an Interstate System highway.

For example, on the East Coast, I-95 and U.S. Route 1 occasionally share the same piece of highway, and the two shields are placed together. As long as the overlap doesn't drag on for too many miles, it's allowed.

## BELTS AND SPURS AND TURNPIKES

To keep things simple, the rules laid out in 1957 said the main Interstate System's highways had to be one- or two-digit numbers. Today they range from I-4 to I-99. The thinking was to stretch the rules and label the beltways around cities with three-digit even beginning numbers and the spurs shooting off the Interstate, running straight into and out of cities, with three-digit odd beginning numbers.

Instead of using entirely new numbers for the beltways and spurs, it was decided that the highway involved just add an even number to the front of it for a beltway or an odd number for a spur. Today, that means a driver in Georgia, traveling on I-85 into Atlanta may drive through the city on I-85 or take Atlanta's beltway, I-285, around the city. While still in Georgia on I-85, the same driver continuing south may take a beeline spur into Columbus, Georgia, labeled I-185.

Belts and spurs aside, the red-white-and-blue shields were approved for posting on Interstate System highways on August 17, 1957. A few days later, the Bureau of Public Roads incorporated 2102 miles of existing toll-charging turnpikes, bridges, and tunnels into the Interstate System, making the toll facilities the first parts of the Interstate System to carry the new red-white-and-blue shields. Bertram Tallamy's New York Thruway was one of the first toll roads to join the System, along with the Pennsylvania Turnpike, a.k.a. the Granddaddy of the Turnpikes.

## TOLL ROADS NEVER DIE

The move made economic sense, since the states had chosen the most heavily traveled routes for their new toll roads and because these routes were serving the corridors between cities that the proposed Interstate System was to service. States could keep their tollbooths up, collecting fares from drivers, until the construction bonds for building the roads were paid off. This saved the states from building a brand-new highway that would compete with their own toll roads.

However, there was a catch. The states wanted to be paid back for the roads that were being incorporated into the Interstate System. Their argument was simple: They had built the roads with their own money and missed the chance to be reimbursed for 90 percent of the cost, while other states were now receiving cash payments for their new highways.

*123*

# The Turnpike That Ended in a Cow Field

**A**MOS SWITZER, an Oklahoma farmer, was a compassionate and peace-loving man. That was until the opening of the Kansas State Turnpike in 1956 converted him into an aggravated agrarian.

Amos had the misfortune of owning a farm on the Oklahoma/Kansas border at the exact location of the new turnpike's dead end. High-speed neophytes who had never seen a superhighway, let alone driven on one, barreled down the newfangled superhighway, overshooting its dead end and crash-landing in Amos's fields. The harried farmer was rescuing the marooned motorists and attending to the bewildered and injured at the alarming rate of one mislaunched car a day. Kindheartedly, he set to plowing his field for softer landings as well as patching the holes in his fences created by the wayward vehicles. But when Wyoming's governor, Millard Simpson, and his wife took the plunge into the fertile fields, bringing both farmer and governor unwanted national fame, Amos was fed up.

The now angry agrarian requested that his neighboring state end the madness by erecting a large barricade at the end of its turnpike. The Kansas Turnpike Authority complied and built an enormous wooden barrier. However, within 24 hours, three cars careened through it, reducing it to splinters.

For better or worse, 18 months and 500 car accidents later, parts of Amos's fields, along with the Kansas Turnpike, were converted into I-35, making the badly needed connection between Oklahoma and the Kansas Turnpike and bringing to an end a distressful chapter in farmer Switzer's life.

It was Amos's own state, Oklahoma, that sparked the idea of a superhighway through his fields. Oklahoma had promoted a plan to build a superhighway from the Great Lakes to the Gulf of Mexico. However, the scheme required that every state along the route agree on the exact locations of the transcontinental highway's connections, raise its own funds, and then construct its own segments of the highway.

In Kansas City, on October 21, 1956, Gene Autry, riding his horse Champion, jumped through a giant

**June 4, 1956:** *Life* magazine's story on why states couldn't build their own road networks.

map of the Kansas Turnpike, heralding the opening of that portion of the multistate highway, and delivering a roadway that met the design standards of the recently announced Interstate System.

But on the day of the Kansas Turnpike's opening festivities, Oklahoma was still only talking about its portion. Plans for the turnpike lagged when the state was unable to raise the money needed for construction, making for the dead end in Amos's fields.

Truncated turnpikes and sudden dead ends had been a recurring phenomenon in the days before the Interstate Decade. Ike's Grand Plan eventually incorporated some of these turnpikes into the Interstate System, as well as helping fiscally challenged states like Oklahoma build the highways they needed. Amos Switzer's story is a textbook case of why the federal government, not the independent states, needed to orchestrate the building and help with the funding of an interstate system. Without that coordinated effort, the country would be left with a hodgepodge of disconnected state turnpikes ending in the cow yards of unlucky farmers.

MISCELLANY

## SUDDEN STOP FOR A HIGHWAY

The new $160 million Kansas Turnpike starts from Kansas City, runs through Topeka, Emporia and Wichita, sweeps majestically across the wide, rolling wheatland—and ends ignominiously at the edge of a contour-plowed field owned by Amos H. Switzer of Braman, Okla. The turnpike, of course, is supposed to keep right on going south to Oklahoma City. But though Kansas has nearly completed the work on its 234 miles, Oklahoma has not yet allotted the money to build its 92 miles and work has stopped at the state border, marked by the road running through the middle of this picture. Until Oklahoma shells out, Farmer Switzer can till his land, safe at the end of a magnificent four-lane highway to nowhere.

Congress's report determining what would be a fair amount to pay back to the states came out in 1958 with some bad news. Over 2.5 billion dollars in payments was due to the states for their toll roads. Since the country was in a full-blown recession at the time, Congress took no action. Hearings were held again on the issue in 1959 and 1966, but still Congress took no action on the study.

Today, we now have tolls on five percent of the Interstate System, and those tollbooths may never come down. Congress has been encouraging states to keep collecting tolls to help with the endless burden of maintenance, repairs, and the expansion of the Interstate System.

## THE RIGHT OF WAY

With signage chosen, building materials identified, and funding in place, a few more hurdles remained to be conquered. Chief among them was choosing the routes for the new highways.

Taking to the sky and flying low, engineers crossed the country, photographing likely paths for the new Interstates. The resulting aerial photographs were then plotted out along with topographical elevations, rivers, other highways, and the locations of towns and cities. The golden rule was: Find the shortest distance between two cities that would be the least expensive to build and would serve the most people. Generally this meant the new highways would not be laid on top of the old state routes, clogged as they were with stoplights, mom-and-pop businesses, and city traffic. The Interstate was meant to be straight, fast, and sleek. If it had to, it would burrow through mountains and cross rivers. And if it had to, it would plow through some farmer's cow field, or bisect his home and his barn.

At this point in the building of the Interstates, there were few objections that turned on the issue of a highway's running too close to a home or business. More often it was the reverse. People objected because the Interstate was bypassing them, streaking through unpopulated areas far from the small-town diners and gas pumps that had served the old routes and provided the populace with a living.

Once a path was chosen, public hearings were held in the various localities and concerns were aired. This was a new experience for the highway engineers, who were accustomed to dealing with geotechnical data, not human emotions. The engineers were ready to explain to a town why a route was well suited in terms of its earth density or the lay of the land. They were less equipped to explain why it should wipe out a family garden or a family's livelihood.

When building I-40 in California, engineers considered using nuclear bombs to vaporize part of the Bristol Mountain range. V.I.P. seating was even planned for the event. The bombing was to produce a cloud 12,000 feet high, and a radioactive blast 133 times more powerful than the bombs dropped on Hiroshima and Nagasaki. Not to worry, the project manager would determine when radioactivity levels were low enough for workers to return to the site and begin building the highway.

*125*

Young engineers of the day got a crash course in public relations. As one highway engineer, Everett Preston of Ohio, put it, "I used to tell our people to always go to the garden club meetings and explain things to them," and his strict advice was never "argue with the chamber of commerce."

Most times, romancing the garden club and buttering up the chamber of commerce was enough. The routes, carefully chosen and explained, were accepted, and rights-of-way were purchased by the government in an amicable negotiation. Most of the earliest highways were in wide-open rural areas. It's a big country, and the highway departments were sensitive to local concerns. Less pleasantly, the right-of-way could be taken by court order in an eminent-domain process—the right of government to seize private property for the benefit of the public, and to require the landowners to accept the terms of the sale.

Often the farmer or homeowner in an Interstate's path was unaware of the true value of his property. When the time came, a knock at the front door from a negotiator representing the state highway department started the bidding. These trained professionals often were able to wangle a sale that was below the market value of the property, leaving the citizen with a bad deal. It saved federal funds, but it wasn't pretty.

On the other hand, some landowners turned the tables on the state and managed to sell their land at inflated prices. To even things out, Congress would eventually pass a uniform-acquisition bill in 1970, ensuring a property owner a fair assessment of the damage done to him by putting four lanes of Interstate through his property.

## AN UPROAR OVER ACCESS

Another thorny issue was access to the new Interstates. In earlier highway-building times, states gave virtually unlimited access to those who lived along the road's path. That was why there were so many traffic lights accommodating side streets. Highways were meant to give people living nearby a way to get to the next town. When a new U.S. route came to a town, large or small, it often became the town's Main Street, open to neighboring dry goods stores and whatever else pleased the citizens.

But the new highways were meant to carry people from state to state or even beyond, not from town to town. Most of the designs called for access points to be three miles apart. A big town might have several access points and a small town none. Even if there was access in a town, it might be on the outskirts, deliberately avoiding downtown congestion.

Limited access was a radical concept, and one that many towns had trouble absorbing. Why should a farmer have to drive three miles out of his way to

The Interstate System is the largest earth moving project in the history of the world. Roughly, 42 billion cubic yards of earth were moved to build the System. The Panama Canal moved a mere 362,000,000 cubic yards.

get on a road to deliver his hay? Why would anyone have to drive to an access point in the next town, when the highway you wanted ran right in front of your house?

Even more important than convenience was economic survival. Engineers could explain that limited access made for a safer ride because there were fewer merges. They could say that limited access meant drivers could safely zip along at high speeds. But the sales pitches all rang hollow to the motel owner whose business was bypassed as motorists hurtled along at those high speeds away from his empty rooms. Limited access had no charms for the downtown department store or drugstore that would see most of its customers diverted to the new Interstate.

Often, though not always, a compromise was reached. Oklahoma was one state that successfully fought the Feds on the route issue. Distraught farmers, businessmen, chambers of commerce, and small-town mayors were convinced that financial doom would be their destiny if the new I-35 was built too far from their existing highway, U.S. Route 77. In a do-or-die campaign, with the help of their governor, Henry L. Bellman, the locals from the towns of Wayne, Paoli, and Wynnewood fought to lay the new Interstate down directly next to, practically on top of, U.S. 77. If the Bureau of Public Roads refused their demands for this route, the governor of Oklahoma threatened to build his own toll road, forgoing the 90 percent federal funding for I-35. State legislators joined the show of solidarity and passed legislation preventing state monies from being spent on I-35 if it did not stay within 5280 feet of Wayne, Paoli, and Wynnewood.

The towns succeeded, and today U.S.-77 acts as a feeder road to I-35. Campaigns like this happened over and over again across the country, especially in rural areas, which were so dependent on highway traffic for survival.

## THE EARTH MOVED

There has never been a project in the world, throughout history, that has moved more earth and shifted the geography of an entire continent as much as the construction of the Interstate System. In 1859 the Suez Canal began its work as the largest earthmoving project as of that time, with over 20,000 workers walking out of the canal's trench with baskets of dirt on their heads, removing nearly 100,000,000 cubic yards of material. In 1904, Americans took over the failing Panama Canal project. Using steam shovels and horses, they spent 10 years removing just over 262,000,000 cubic yards of dirt, creating Teddy Roosevelt's "Big Ditch."

In 1956, the United States began removing an estimated 42 billion cubic yards of earth to make way for the Interstate System. A large portion of that

# Breaking Barriers

**B**ETTY ANDERSON, TV's favorite daughter in the 1950s, had an ambition she couldn't keep to herself. Nervous and proud at the same time, she blurts out her secret to her parents. "Mother, Father, I have something to tell you . . . something quite important! I am going to be an engineer!"

The show's background music mimics a heavy "uh-oh" as the parents on the prime-time television show *Father Knows Best* can't hide their dismay. It was April 11, 1956, and women were definitely not welcome in the all-male world of the engineer.

Betty does wangle her way onto a highway surveying crew, but she soon surrenders her ambitions, and winds up going out on a date with a young man from the crew. A prudent move, since father and the young man have already agreed that having Betty working on the highway would be most inappropriate. The men don't want their women out surveying highways. Every guy on the road crew just wants to "come home to a nice pretty wife."

Commenting on the show, Robert Thompson, a professor of media at Syracuse University, says, "It's almost as if she's foreshadowing a whole new way of doing things." But she is just too far ahead of the times, he says. The show "ultimately has to be very protective of the status quo. . . . Betty had to be put down."

In 1956, a woman engineer was unthinkable. But as the Interstate System gathered steam, the nation's highways helped women onto their own roads to success.

## "THE BUREAU'S PRETTIEST ENGINEER"

On April 2, 1962, flashbulbs were popping and cameras snapping as five top officials from the Bureau of Public Roads welcomed their newest engineer. Not every new hire got this kind of treatment; but this engineer was a she, not a he, and therefore all the fuss. The men, old enough to be her father, couldn't contain themselves. Like boys from an all-boys school seeing a girl in their midst, they were gushing.

Miss Beverly Cover, twenty-two years old and just out of Georgia Tech, was the Bureau of Public Roads' first female engineer. Calling her a "Cover girl," naturally, and the "Bureau's Prettiest Engineer," publications of the day reported that she was not only the second female engineer ever to graduate from Georgia Tech, but was also a "baton-twirling majorette."

"Miss Cover sees nothing particularly strange about a pretty young girl being a traffic engineer," wrote *The News in Public Roads,* a monthly paper out of the United States Department of Commerce. The chief of the Traffic Operations Division, the head of her new workplace, said that before the "pretty young girl" showed up to work, everybody in his division complained of a lack of office space. "Now," he said, "everyone has suddenly discovered that he has room in his office for another desk."

Cover, though, lasted only two years at that desk before calling it quits "in order to become a full-time housewife and mother," as *The News in Public Roads* reported.

## "GET THOSE WOMEN OUT OF HERE!"

Women may have been heartily welcomed in the offices of the Bureau of Public Roads, but when it came to the job sites themselves, it was another story.

The Eisenhower Tunnel, high in the Rocky Mountains and part of I-70 as it passes through, not over, the Continental Divide, proved to be a mind-altering experience for some of the crews building it. Not only because it took 16 years to bore their way through 1.6 miles of rock at well over a mile above sea level. But also because a woman named Janet Bonnema showed up to work along side them.

In 1970, Janet Bonnema applied for a job with the Colorado Highway Department. After taking and achieving a very high score on the application exam, Bonnema received a letter offering "Mr. Janet P. Bonnema" a position as an engineering technician at the Eisenhower Tunnel, which was then under construction. When she checked on the offer, a state employment officer nearly begged her not to accept the slot saying, "Women are taboo in the mines and tunnels of Colorado. Those workers would flat walk out of that there tunnel and they'd never come back." Nevertheless, "Mr. Janet Bonnema" showed up for work.

The astonished crew at the site was forced to accept her, but she was not allowed to enter the tunnel. Citing the workers' foul language, inappropriate toilet facilities, and folklore, the state gave Bonnema a desk job instead. The woman—who was a rock climber, an alpine skier, and an airplane pilot—was steamed. "I am not allowed to do the same work as the male engineering technicians even though I am physically able, in better condition, and have more stamina than many of the male engineering technicians," she complained.

In November of 1972, after 18 months of sitting at a desk, a change in the state's equal employment laws, and a court order, Bonnema was finally allowed into the tunnel. Accompanied by a woman reporter covering the historic moment, she made her way into what construction crews called the Big Hole.

"Get those women out of here!" yelled one worker. Rumors and superstition ran rampant. Another worker was quoted as saying, "Some years ago I took my wife into a tunnel and the next day a man got killed."

Undeterred, Bonnema trudged into the muck of the tunnel, triggering a walkout of over 60 workers. One of them turned in his gear and quit on the spot saying, "They had a woman in the tunnel, and I will not work there for that reason. It's a jinx. I've seen too many die after a woman was in the tunnel." Ms Bonnema's response was simply, "They're making $8 to $10 an hour. What's the matter with them?"

The walkout cost the state about $10,000 in lost time and wages, but the next day the men returned and construction resumed. Bonnema stayed on to work beside them. The tunnel was successfully completed and is still in operation, with no unusual mishaps, despite the presence of a woman on its construction crew.

On March 8, 1973, when the first tunnel was opened, it was a woman who had the last word. Mamie Eisenhower wrote a letter to Colorado's governor John Love, in recognition of the fact that the engineering marvel was named after her husband. "How proud Ike would have been with this honor," she said, "for he loved Denver and Colorado."

## FROM SECRETARY TO THE CHIEF

Mary Peters packaged the meat of hogs when she was a teenager. When she was old enough to drive, she rode them down the highway—the Harley-Davidson species of "hog" this time. Now she is in charge of many of the roads on which she once drove her

**Mary Peters,** Chief of the Federal Highway Administration.

motorcycle. In a meteoric rise to the top, she has completed the circle for women in the highway engineering business.

In 1985, after raising three children, Peters took a job as a secretary at the Arizona Department of Transportation. Within six months she was promoted, beginning a steady climb to the number one spot as director of the department. In 1998 she was the "man in charge" down in Arizona when she got a call from the White House asking her to head up the Federal Highway Administration, the first woman to fill that role.

Over 80 years after Chief MacDonald first took over the Bureau of Public Roads and nearly 45 years after Ike inaugurated the construction of the Interstate System, Mary Peters began leading the way on the nation's highways. She improved security along the country's borders in the wake of September 11, helped the military mobilize for the second Gulf War, and transitioned the Interstate System program from construction to maintenance and improvement. With her arrival in Washington, women can at last say they have gone the distance in building the nation's highways.

*When Ike inked the deal to create the Interstate System in 1956, he virtually launched a new industry, the manufacture of heavy earthmoving equipment.*

Enough concrete was used to construct the Interstate System to build a wall nine feet thick and fifty feet high around the world.

earth was moved in the first 10 years of its construction.

When Ike inked the deal to create the Interstate System in 1956, he virtually launched a new industry, the manufacture of heavy earthmoving equipment. Before World War I, steam shovels typically removed rock and earth for a road's excavation, with men shoveling the stuff into wooden horse-drawn wagons. After the war, diesel-driven trucks and bulldozers began making their way onto highway projects. Not until 1932, when Caterpillar Inc. began painting its machines "Hi-way Yellow," separating itself from manufacturers of farm equipment, did the industry begin to evolve. But the really big gear was designed and built solely for the construction of the Interstate System.

The years between 1958 and 1963 were consumed with digging the roadbed for the new superhighway. Five key manufacturers engineered and assembled the behemoth contraptions that did most of the digging. Caterpillar, International Harvester, Allis-Chalmers, LeTournou, and General Motors churned out massive scrapers that were capable of picking up the equivalent of over 1000 steam-shovel loads of earth in a single bite. The scrapers delivered the earth to powerful bulldozers that pushed the dirt with new high-tech, high-strength blades. Dump trucks, so large they were illegal to drive on the public highways they were building, were fabricated to haul loads equivalent to over 40 horse-drawn wagons in a single trip.

Somehow, building the highways became routine.

To begin, general contractors unleashed their excavators and began clearing and grubbing the job site. A crude path about 200 feet wide and several miles long was staked out and then cleared through the woods and farmland. Half of that 200 feet was to accommodate four lanes of highway, two lanes in each direction, each requiring 12 feet in width; a median at least 36 feet wide; a pair of 10-foot-wide breakdown lanes, one in each direction; and a pair of three-foot-wide left-hand shoulders. The other half of the 200 feet was for additional right-of-way along the sides of the road, for utilities and for possible future expansion of the System.

Teams with axes and relatively new tools called chain saws cut down the trees. Bulldozers followed them, stripping the area of topsoil. Large backhoes unearthed the tree trunks, ripping up their roots. This organic refuse was either bulldozed into deep holes and then buried or gathered into a large pile and burned. Today, these valuable by-products of construction are stockpiled and reused. Wood chippers turn branches and tree trunks into mulch for the highway's final landscaping, and valuable topsoils are saved to set in new plantings along the roadway.

In the hilly terrains, a rhythm of cutting and filling developed. The best route, from a contractor's point of view, allowed for an even amount of cuts and fills. In other words, every time a huge cut in the mountainous terrain was made, an equally large area of filling would be required. Taking material removed from the mountain cut, the contractor built the new road's approaches with the fill. In a perfect situation, this reduces the need to haul material away when digging through a mountainside, and it prevents the contractor from having to haul in material from great distances for filling operations.

After the clearing and grubbing and the cutting and filling were complete, the contractors called in the excavators, who used some of the world's largest earthmovers, called scrapers. Dragging their bellies along the ground, usually not far from the new highway's path, these behemoths scraped up to 90 tons of earth at a time. With their cargo of dirt, the scrapers lumbered over to the roadbed, where the operator gradually released the load of fill along the highway's alignment, slowly building the roadbed up to within inches of the engineering drawings' specifications.

After a scraper unloaded its shipment of dirt, it returned to a designated area to take another bite out of the ground. Over time, this created a deep hole, called a borrow pit. Acting like on-site quarries, the borrow pits supplied the

**One of the largest** machines ever built for the construction of the System is seen here loading 3000 cubic yards of earth an hour into waiting scrapers.

# The Interstate as Archeological Dig

**O**NE HUNDRED AND FORTY MILLION years ago, while grazing on marsh grasses along the subtropical Sundance Sea, a very large and slow-witted brontosaurus took a step he would deeply regret. He planted his giant foot in a marsh too soft to hold up his 50 tons of meat, bones, and fat and, while wondering what to do about it, the monster slowly sank until he was gone.

Flash forward 140 million years, to 1955, as a search team scouring the earth for uranium responds to the squelching and buzzing of their Geiger counter. Radioactive emanation from the dinosaur's bones has triggered the team's sensitive electronic sensors. The group begins digging. Disappointed to find fossilized bones and not uranium they move on, but not without marking the site for posterity.

Enter paleontologist Mike Mayfield and plans for the Interstate System's I-70 superhighway near New Castle, Colorado. Fifteen years after the uranium fortune hunters, Mayfield and his crew, along with some heavy equipment from the state highway department, began digging, chipping and extracting the fossilized brontosaurus, which happened to lay in the path of I-70 between Denver and Grand Junction Colorado.

This was just one of hundreds of digs conducted along the Interstate System, a link from the very distant past to the present. The careful treatment is the result of The Historic Preservation Act, legislation which requires Interstate builders to avoid areas of historic importance if possible, or if not, to treat them sensitively.

## THE VERTEBRAE IN THE FIREPLACE

Chiseling away at the lighter rock around the darker rock that are the fossilized bones, the team wrapped their find in plaster, removing it for future studies. They even made a house call to a local man, taking plaster casts of three brontosaurus vertebrae that were imbedded into his fireplace as decoration. Mayfield said he wouldn't ask for them back. "He was good enough to tell me about them, and besides the bones aren't the primary object. We want to know what kind of animal he was; why he was here, what the environment was. Did they kill each other off? Did they overburden their environment? We need to know."

An open house of the dig site drew nearly 2000 motorists, schoolchildren, and tourists from 28 states, who stopped in to pay their respects to the brontosaurus and see for themselves what was about to be paved over. A small accident on I-70 snarled traffic and prevented even more people from coming.

But then if there hadn't been an accident 140,000,000 years ago, there wouldn't have been an open house at all.

**Mike Mayfield,** a paleontologist, holds a fossilized bone of a 140,000,000-year-old, 50-ton dinosaur that was removed to make way for I-70.

road builder with all the earthen material needed to create the roadbed. Today, when driving along the Interstate System, you can occasionally see these borrow pits pockmarking the landscape. After water found its way into them, they became large ponds or small lakes.

Since the building of roadways in ancient Rome, water has been the demise of many highways. To whisk the liquid away from a road's foundation, engineers designed elaborate drainage systems. Tens of thousands of miles of drainage pipes were laid, crisscrossing beneath the Interstate System's 41,000-plus miles of highways and entombed in the gravel and concrete beneath the highway's thick subsurface.

Once the embankments and the drainage systems were in place, the road's pavement structure was built from the bottom up.

The pavement structure is the layers beneath the road surface that support and distribute the weight of the traffic and prevent damage to the road surface. Designs for the road vary from one state to another, based on weather conditions and available material, much of which was determined by the test track in Ottawa, Illinois.

First, dump trucks poured a thick layer of large gravel stones onto the roadbed, making the subbase or foundation of the paving surface. On top of the subbase, a thin layer of hot liquid asphalt was poured to prevent ground and surface water from creeping into the highway's foundation. On top of this thin coating was the roadway's base and final travel surface, a layer of concrete poured around steel bars called reinforcement bars or just rebar. Thousands of miles of rebar, tied together by hand with pliers, wires, and wire cutters, were needed for each mile of superhighway.

The depth of the roadway varied according to its location. In remote desert areas, where traffic loads are light and weather is not a major factor, the highway might be less than a foot thick. In Maine, the brutal winters and frost required that sections of I-95 be up to five feet thick. Starting from the bottom, a 24-inch layer, or subsurface, of sand and gravel was dumped into the Maine superhighway roadbed, followed by 18 inches of a coarse gravel and a five-inch layer of crushed and processed gravel, and finally nine inches of surface pavement.

The Interstate System mostly had a concrete surface during the early building. Improvements in asphalt, which have made it a low-cost, easy-to-use material, didn't come about for another 20 years.

After about a month of curing, the concrete was hard enough to allow cars and trucks to drive over it. Finally, the landscaping, painting of lines, and placing of signs along the road were completed and the obligatory ribbon cutting was performed. Eager Americans wasted no time putting the road to use.

*In remote desert areas, where traffic loads are light and weather is not a major factor, the highway might be less than a foot thick. In Maine, the brutal winters and frost required that sections of I-95 be up to five feet thick.*

133

**An entire heavy earthmoving equipment industry** came into being as a result of the Interstate System. *Above:* A scraper delivers a 120,000-pound-load of earth, shaking free its payload so the waiting bulldozer can level the Interstate's future roadway to within inches of its required grade. *Inset:* The completed grading of an interchange on I-94 in North Dakota. *Opposite:* The Interstate System consumed over a billion pounds of explosives, three million feet of lumber, 2.3 billion tons of cement and crushed rock, and millions of miles of steel reinforcement bars.

Slowly the miles through America's countryside were laid out, linking one unfinished section of Interstate System to another.

## MILLION-DOLLAR MILES

In the early 1920s, the average mile of highway cost $17,000. By the late 1950s, the average stretch of rural Interstate System highway cost a million dollars a mile. And those million-dollar miles, which might require the moving of over a million cubic yards of earth for each mile built, were the easy part of the Interstate System. In the country, labor and land were cheaper than in the city, and contractors were able to use larger pieces of equipment, reducing the costs of construction by increasing hauling loads. The building pace was fast forward—build and pave, build and pave.

**An Interstate System** paving train: Processing raw concrete at the front of the train, it transforms it into a finished superhighway surface at the rear at the rate of 600 feet an hour.

President John F. Kennedy, in one of his final acts, cuts the ribbon opening the Delaware–Maryland Turnpike on November 14, 1963. The President told the assembled crowd of 10,000 on the cold day that the highway "symbolizes the effort we have made to achieve the most modern Interstate System in the world, a system which, when completed, will save over 8,000 lives a year." This turnpike was the last one allowed into the Interstate System when the Federal Government said the two states could forgo the standard 90 percent reimbursement and collect tolls instead. The road has since been renamed the John F. Kennedy Memorial Highway. Today, its 53 miles of toll road are the most expensive to travel on, per mile, on the Interstate System.

While the state highway departments were moving quickly, so were Americans, who took to their new highway system with great enthusiasm . . . and speed. At the end of the Interstate Decade, posted speed limits of 70 miles per hour were common in Iowa, South Dakota, and Nebraska. In Kansas, along a stretch of I-35, 80 miles per hour was the posted limit; and in Montana the speed limit was set at "reasonable and prudent." Americans were becoming addicted to speed.

Work progressed so quickly during the first 10 years that the supply of raw materials the project needed began drying up. Steel, sand, and cement became more difficult and more expensive to find as billions of tons of the stuff were consumed. Even the supply of young male engineers was tapped to its fullest, and for the first time women were hired by state engineering departments to help design highways. Innovations developed so often that it was a challenge for highway departments to keep up with new techniques. Inside the Kansas state highway department, the responsibility for getting engineering and construction news to the men in the field fell to "circuit riders." Think of Paul Revere in a pickup truck and you have a pretty clear picture of what went on. Dispatched from headquarters with information on an ingenious use of materials discovered by another state, circuit riders drove from one field office to another. Arriving at a construction site, the circuit rider sought out the lead field engineers and relayed the information. Once that was done, the circuit rider was off to the next site.

The states also began work on the more difficult Interstate miles, those that ran through the heart of the country's urban centers. The urban miles demanded the most lead time in planning, approvals, and purchasing rights-of-way. Just 5201 miles of the original 41,000 miles of Interstate highways would be built in urban areas, but those miles would drive about 50 percent of the cost. Still, no one really knew what to expect when building in America's downtown.

CHAPTER EIGHT

# The Interstate Goes Downtown

I N THE SUMMER OF 1959, President Eisenhower and his entourage found themselves stuck in a monumental traffic jam while en route to Camp David, the presidential retreat outside of Washington, D.C., now named for his grandson, David Eisenhower. Ike, who could be short of patience and temper, demanded an explanation for the delay. He was dismayed to discover that the cause of the traffic snarl was his own Grand Plan. Blocking the presidential path were construction crews building an Interstate System directly through a suburb of the nation's capital.

Superhighways so close in to the city? This is not what the President had in mind when he launched the Grand Plan. To make matters worse, he would soon discover there was an Interstate System highway planned to pass beneath the reflecting pool in back of the Capitol.

Angered by what he saw as a move "against his wishes," Ike launched a presidential inquiry to advise him on what had gone wrong. He found out that the problems could be traced to a powerful set of documents bound between covers and called the Yellow Book.

## THE YELLOW BOOK
If Dorothy and her friends followed the Yellow Brick Road to Oz, The Interstate System and its highways followed the Yellow

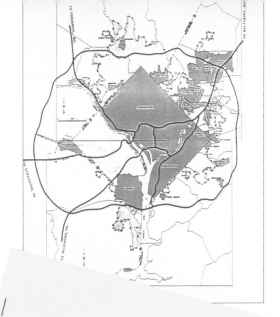

General Location of
National System of Interstate Highways

INCLUDING ALL ADDITIONAL ROUTES
AT URBAN AREAS
DESIGNATED IN SEPTEMBER 1955

U. S. DEPARTMENT OF COMMERCE • Bureau of Public Roads

For Sale by the Superintendent of Documents
U. S. Government Printing Office, Washington 25, D. C. Price 55 cents

**The Yellow Book's** 100 pages detailed the path of the Interstate System through 100 American cities. It promised that a vote for Ike's Grand Plan would be a vote for massive expenditures in each of the listed cities.

Book through America's downtowns. Officially named "The General Location of National System of Interstate Highways: Including All Additional Routes at Urban Areas," the Yellow Book was a 100-page document that was delivered to every Congressman's desk in September of 1955. It documented the crude paths of the soon-to-be-constructed Interstate System through the lawmakers' hometowns, with simple maps of 122 cities in 43 different states. When the vote for Ike's Grand Plan came before the U.S. Congress in 1956, the Yellow Book played a significant role in getting Eisenhower the support he needed.

It was the Bureau of Public Roads, the old bailiwick of Chief MacDonald, that had created The Yellow Book, in conjunction with state and city governments. The document informed elected officials of the urban miles of federally funded Interstate System that would be built if they voted for Ike's legislation. Every Senator and Representative with a city center in his district would see millions of dollars in federal funds flow in, bringing desperately needed jobs and new infrastructures. The Yellow Book went a long way in securing congressional support for the Federal Aid Highway Act of 1956. MacDonald's belief that the cities must be served was a major influence on the final path of the Interstate System. The Chief was gone, but not forgotten.

One of the strangest things about all this is that Ike knew nothing about the Yellow Book. His staff, he later said, had told him it was a historical narrative of highway legislation. In reality it contained over 100 black-and-white maps of urban highways Ike never wanted to build.

How could Ike have been unaware of such a major change to his plan? Very easily, considering that in 1955–56 he was also dealing with the threat of a nuclear holocaust and the Cold War, not to mention the fact that he was fighting for his life, recovering from a heart attack and intestinal ailments.

Ike was an astute politician who knew how to choose his battles. He was nearing the end of his presidency when he learned of the Yellow Book and its direct contradiction of his wishes. But this was a battle he chose not to fight. The United States Congress had voted in favor of building in the cities and he was not about to take them to task. His main concern was keeping the Grand Plan on course. If building through the downtowns ballooned the budget beyond all expectations (by 1960 the costs were already approaching 40 billion dollars) that was a matter for the next administration to deal with. . . . His hands were tied.

The Federal Highway Administration suggests that states place exits on the Interstate System every three miles. In metropolitan areas they suggest, but do not require, an exit each mile.

## SPONTANEOUS COMBUSTION

By the time Ike made his decision to let the Interstate System continue on through America's downtowns, many of the nation's cities were in tough shape. For nearly three decades the country's urban cores had been experiencing an economic, social, and cultural drain. For years families and businesses had been packing up their cars and trucks and heading to the suburbs, moves made possible by the newly built Interstates. Left behind were downtowns with a hollow feeling, decaying houses, and neglected parks. Because more and more people were driving, the narrow streets and alleys in and around the cities' central business districts were choking on traffic, jammed with more trucks and cars than ever. Tax bases were declining and the quality of urban life was hitting rock bottom. Governors, mayors, and every other elected official representing a downtown constituency were hungry for investments to stop the downward spiral. The Grand Plan seemed to be the answer.

The cities' pent-up demand for investment and the billions of dollars available from Interstate funds made for a spontaneous combustion of construction around the country. Every major city seemed to have an unrealized master plan for an expressway. The 90 percent federal funding meant they could dust off those old blueprints and begin implementing them, as long as they were in the Yellow Book.

At first, urbanites and suburbanites alike wanted to see the metropolitan sections of Interstates built. Those who lived in the city wanted a job, relief from congestion, and some hope that their cities would improve. Many of those who had left town for the suburbs were still commuting back to their offices in downtown. Cities were still the centers of transportation by sea, rail, and air, and the new expressways would make those connections easier. It looked as if the urban Interstate System highways would be the panacea for inner-city ills.

When launching the Interstate System, John Volpe, President Eisenhower's first federal highway administrator, called the nation's top state highway officials together for a meeting in Atlantic City, New Jersey. Get going on those

*The cities' pent-up demand for investment and the billions of dollars available from Interstate funds made for a spontaneous combustion of construction around the country.*

*141*

highways as soon as possible, he told the states to think and act big.

And Volpe had an agenda. He asked the state officials to build the urban sections of their Interstates first, urging them to start the program in their downtowns where traffic demands were heaviest. He also warned them not to let the highways they were about to build destroy the communities they were meant to serve. It was a nearly impossible task: build superhighways through tightly knit neighborhoods, but don't ruin their charm.

As the nation's top highway official, John Volpe knew better than anyone just how difficult building a superhighway through the heart of a city was going to be. The former commissioner of the Department of Public Works in Massachusetts, he had just settled one of the nation's first major urban superhighway uprisings, in Boston.

## A HIGHWAY REVOLUTION

Massachusetts, Volpe's home state, had gotten a head start on its road building, preempting the federal Interstate program. Not waiting for the federal government to fund its highways, Massachusetts had taken a 1948 Master Plan and, with its own money, begun construction for an elevated highway through its state capital, Boston. The project was a favorite of Volpe's predecessor at the DPW, a tough, house-bulldozing, highway-paving kind of a guy named William Callahan. Those not fond of Callahan's bully tactics and hardball politics called him Frankenstein. More neutral folk referred to him as the Highway Czar.

Callahan's blueprint for building highways in Massachusetts was one of the country's first statewide highway master plans. It was an ambitious project calling for nine arterial expressways. Reinforcing Boston's nickname, "the hub," the expressways were to emanate from Boston's center like spokes from the hub of a wooden wagon wheel.

Eagerly, Callahan began gearing up for construction on the scheme's most essential component, the Central Artery, his "Highway in the Skies." In the era of Buck Rogers and Flash Gordon, the three miles of steel and asphalt were to whiz traffic over the city, conveniently passing over neighborhood rooftops at heights of nearly 80 feet. The intent was to attract motorists into the city, allow them to pass over the congested streets below, and drop in and out of the city on any one of the Central Artery's 34 off- and on-ramps. Suburbanites would go to their places of work, to their favorite department stores, or to train stations and then, zip! they would be back up on the Highway in the Skies.

In 1950 the DPW began building an underwhelming bridge, one only an engineer could love, at the Charles River, just north of the city. The following

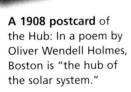

**A 1908 postcard** of the Hub: In a poem by Oliver Wendell Holmes, Boston is "the hub of the solar system."

*Above:*
**Every North End building** in this panorama was leveled to make way for Callahan's "Highway in the Skies."

*Left:*
**Pushing the plunger,** Callahan said, "I only wish my critics and enemies were sitting on top of that ledge."

*Boston Sunday Globe* ROTO *Pictorial* magazine
AUGUST 8, 1954

**Boston's $110,000,000 Highway in the Skies**

year the DPW continued on its way into Boston, ripping away at an ethnic neighborhood called the North End and a historic district called The Bulfinch Triangle. And there was more of the same coming for the rest of the city, as this was only the first of over nine major projects on Callahan's drawing board.

No sooner had construction begun than the sparks of controversy started flying. Bostonians watched in horror as people were forced to leave the homes they'd lived in for generations and as their historic neighborhoods were desecrated. Bucking a national trend, Bostonians began rising up against a highway through their city. In small pockets, then in larger and louder ones, the people

*Above*:
**Volpe's Tunnel** through Chinatown.

*Opposite*:
**By October 1954,** Boston's Central Artery was blasting its way through Boston's North End, creating a 40-foot-high, 100-foot-wide swath through one of the densest neighborhoods in the city. Over 1000 buildings were destroyed—many of them historic. Astonishingly, the open-air food market in the far left of the inset photo, opposite, survived and to this day thrives.

let their voices be heard. America's highway revolution, for better and for worse, had begun.

In 1952, bowing to the will of the people, the state stopped construction of the elevated highway just before it entered the city's Chinatown neighborhood, but only after it had torn a wide path through Boston's major Irish and Italian enclaves.

Interestingly, the highway revolution had begun at nearly the exact location as one of America's most legendary revolts. The truncated elevated highway stopped at the point where, on the evening of December 16, 1773, Sam Adams, Paul Revere, and a band of colonists thinly disguised as American Indians, had themselves a riotous Tea Party. Marching from the Old South Meeting House in the center of Boston to the city's harbor, the revolutionists would have passed under the future Central Artery on their way to Griffin's Wharf.

Now, painted green and looking ugly and mean, the Central Artery hung there, poised to plow into the next neighborhood in its path. It stayed there a long time. The standoff between the state's DPW and the city of Boston over the elevated highway had been festering for over a

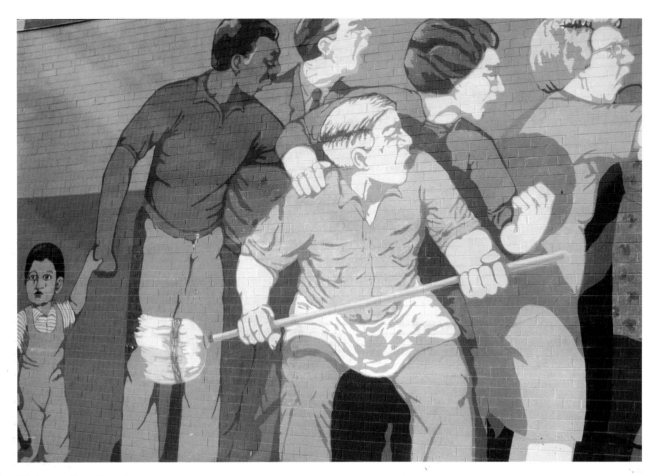

**Highway Revolt** wall mural: In 1969, I-95 was about to rip its way through the heart of 18 Boston neighborhoods. During a tumultuous period, ethnically diverse neighborhoods found common ground fighting I-95 and its beltway, I-695. The neighborhoods won, and today a rapid-transit line operates in the former Interstate System right of way.

year when Volpe stepped in as Massachusetts's commissioner for the Department of Public Works, replacing Callahan. Volpe was empathetic to the plight of ethnic neighborhood and business communities, having grown up the son of poor Italian immigrants. He searched for ways to lessen the long-term adverse effects of a highway on the communities around it. His solution: the largest superhighway tunnel ever built.

Volpe's plan was both brilliant and innovative. He instructed the DPW to build a shallow six-lane tunnel with curves and off- and on-ramps. The modern underpass would have far less negative impact on Chinatown than an elevated superstructure, but its path would be identical to the one the elevated highway was supposed to follow. It would be the first highway tunnel in the world to have off- and on-ramps in addition to its main entrances. Its $18,000,000 price tag was hefty, three times what an elevated highway would have cost. Volpe, however, was passionate about sparing Chinatown and its future generations the blight and division the other neighborhoods had already suffered.

In 1956, just months before Volpe left to become Ike's highway chief, the tunnel's construction began. It was a pioneering decision with more vision than anyone, including Volpe, could ever have realized. In the cities, tunnels were to become a powerful source of mitigation, a form of minimizing the ripping effect urban Interstate System highways have on neighborhoods. American cities

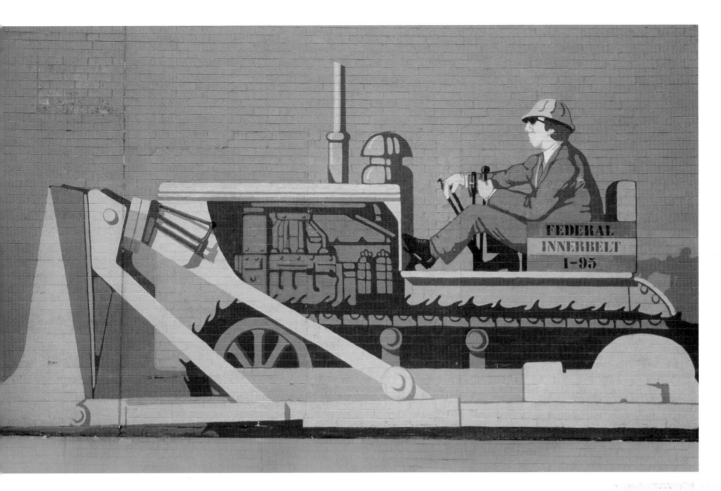

would now be able to benefit from the superhighway while keeping intact the neighborhood life above. In effect, they could have their cake and eat it too.

## COMMON GROUND

In the 1960s, as the highway-building fever gripped the nation, Boston, like the rest of the country, was suffering from the upheavals of the times. With one of the highest populations of college and university students in the country, the city was hosting more than its share of Vietnam War protests. Boston was still mourning the loss of its favorite son, President John F. Kennedy, when a few years later another violent act took his brother Bobby. Racial tensions in the city reached a flash point when a graduate of Boston University's school of theology, Dr. Martin Luther King, was gunned down in Memphis, Tennessee.

Strangely, the one force galvanizing the city during this tumultuous period was the opposition to the bulldozing highways. Ike's grand plan, or Congress's version of it, was in full swing; and a stretch of new highway, I-95, was again slated to run straight through the heart of the city. Its beltway, I-695, would create an iron ring around the city's downtown. The new highway would be a partially elevated structure about 10 miles long, nearly four stories high, with 13 interchanges plowing through 18 heavily populated neighborhoods. In concept

When looking for the quickest way to get to the Interstate from any of America's downtowns, follow small shields called, "Trail-Blazers." These are miniature red white and blue Interstate System shields, complete with arrows and the number of the closest route. Just follow the "Trail Blazers" to the Interstate.

# America's First Beltway

**W**HILE RESENTMENT FESTERED over the Green Monster in downtown Boston, all around the city the glowing success of a new road was taking hold. The road quickly turned out to be as much of a prize as the Green Monster was a failure. What's interesting is that the route started out as something entirely different from what it became, the nation's first beltway.

Like other Americans, Bostonians took to their automobiles and were actively exploring the countryside in the early 1920s. These Sunday drivers were causing heavy traffic congestion along the rural roads leading out of town. To help keep people on course, the state decided it would be a good idea to mark the route of an enjoyable drive sweeping around the city and through the hinterlands. And so a group of independent roads was cobbled into one route, from the sandy beaches of Boston's South Shore through swampy marshlands, pig and chicken farms, finally ending up at the fishing villages on the city's rugged and rocky North Shore. Without making a single road improvement, the state had successfully assembled a chain of recreational outlets. With a leftover number unassigned to any other roadway, officials labeled the semicircular road Route 128.

After World War II, many Bostonians were not content with a Sunday drive out of the city. They were ready to get in their cars and move out to the country for good, and so it seemed were Americans across the country. This wasn't lost on William Callahan, the eager road-building Highway Czar. Plans were already underway to improve Route 128, but under Callahan they went on fast forward. He wanted to build an unheard-of six lanes and was lucky to get four approved.

At first people were skeptical, calling the idea of a highway around the city and through farmland "a road to nowhere" and "Callahan's Folly." But where others saw failure, Callahan saw only potential. He predicted, "This new highway will cause the relocation of business establishments and open new residential sections." And how right he was.

Ground was broken for Callahan's lanes in 1950 and within 18 months the first 23-mile segment was opened. Defying the Bureau of Public Roads' pessimistic prediction that the road would be underused, the gleeful Highway Czar watched as over 76,399 vehicles took trips down the newly poured surface in the first three days of the new and improved Route 128.

At this point the term beltway was not even in use. Officials instead called the road the Great Circumferential Highway. Others would call it the Golden Semi Circle, the Golden Horseshoe, the Miracle Mile, and the Golden Arch.

Whatever bad names they wrestled with, the highway to nowhere kept outperforming expectations. By the 1970s, over 10 times the number of vehicles trips originally predicted for that date were being made along the trend-setting beltway.

## PIG FARMS TO OFFICE PARKS

Years before the highway's construction began, Gerald Blakely was awoken by a thunderbolt of an idea about the coming of Route 128. Slipping into the bathroom to avoid waking his sleeping wife, Blakely grabbed a pad of paper and began scribbling down a corporate recipe that would bring him and the company that had just hired him, Cabot, Cabot, and Forbes, financial success beyond anyone's hopes and dreams. The fruits of that recipe would be Howard Johnson's, Holiday Inn, McDonald's, and the rest of corporate America.

The logic in hindsight is obvious, but at a time when the areas outside Boston were mostly pig farms and meadows, it was a daring strategy. Blakely's epiphany started his firm on a buying spree of large tracts of land at the planned interchanges of Route 128. The firm then applied to change the local zoning ordinances, allowing Blakely and his associates to develop the land for uses never seen in those parts: research, manufacturing, and distribution. Since many of the companies would want easy access to downtown Boston, being near off- and on-ramps was essential.

**In 1955,** Volpe (second from left) opened a segment of the new beltway. To the left is a modern overpass over Route 128 in 1931.

The formula worked so well, Blakely's tactics earned his company the name "Cabot, Cabot and Grab-It." They must have laughed all the way to the bank.

The land was cheap, the traffic was light, and the air was fresh. Companies began moving out of their old, crowded, and expensive downtowns and into new offices built for them by Cabot, Cabot, and Forbes and a slew of other developers. With the companies came jobs, with the jobs came workers, and with the workers came houses. In 1955, 53 companies were positioned along the new beltway. In just 24 months there were 223 companies, and by 1967, 729 businesses called Route 128 home. During this same period, the population of Burlington, formerly a sleepy small town, exploded by eightfold. In some cases, land values went from $450 an acre to over $20,000.

The beltway boom continued in fits and starts through the '70s, '80s, and '90s as manufacturers made room for research and development labs and high-tech firms. It thrived during the dot-coms and survived the dot-bombs. Polaroid, Sylvania, Cleveite, and Canada Dry were among the early alumni, followed by Wang Computers, Digital Equipment Corporation, and Lotus Software Corporation. Large government contractors like Raytheon, AVCO, and a host of smaller ones arrived, drawn by the relatively inexpensive space. The beltway fed the synergy between Boston's businesses and its prestigious universities with their talent pools of faculty and graduates.

The Interstate System's beltways, like Route 128, allowed the cities they served to expand their influence well beyond their previous city lines. Beltways put muscle into an economic expansion that benefited every town and city near them. Route 128 and other beltways around the country put affordable homes within reach of lower-income families. Acting as a decongestant to the crowded downtowns, the beltways spurred on unprecedented economic growth, fostering the creation of start-up companies unable to afford downtown's rents. Many of these companies eventually grew strong and remained loyal to the area, keeping their jobs and goodwill in the community.

The success of America's first beltway inspired the officials running the bad-name game for the remarkable road. Contending titles were the Space Highway, the Electronic Highway, the Technology Highway and, perhaps the most appropriate, the Miracle Highway.

In 1955, Route 128 officially became part of the Interstate System; and by 1974, it was part of I-95. But mostly it's still known around the country as Route 128.

*149*

**The Central Artery,** at center, was slated to be I-95, and the stub at the lower left was to be part of its beltway, I-695, a.k.a. the Innerbelt. With their city already physically divided and damaged by highways, Bostonians killed both projects.

it was a miniature beltway encircling the business district, a noose around the neck of Boston. It seemed no lessons had been learned from the bulldozing '50s.

Governor Frank Sargent, the business community, and the state's power brokers were in favor of the superhighway's construction. John Volpe, the former protector of the neighborhoods, had returned to Washington, D.C., now President Nixon's Secretary of Transportation. From his powerful perch, he too supported the construction.

In fact, the building of both roads was in progress. Along the path of I-95, hundreds of homes had already been taken and demolished in a working-class Irish neighborhood called Jamaica Plain and in a predominantly black neighborhood known as Roxbury. Other sections of the city and a nature preserve awaited a similar fate.

But the obstructionists were gaining in strength and numbers. Professors from Harvard and M.I.T., a lawyer from the predominantly Jewish suburb of Brookline, a group called Boston Urban Priests, housewives from the League of Women Voters, wage earners, and professionals were joining together to oppose this new, intrusive Interstate highway.

First the working-class neighborhoods in Cambridge started resisting only the idea of the highway's path, not its construction. When the DPW refused to discuss the concept of a tunnel or even simply changing the proposed route for a less disruptive one, the community dug in and went for broke. They took up a no-highway-at-all stance. Other communities began to follow their lead. Eventually even the city's mayor, Kevin White, was speaking out against the highways. It was enough to make the man in charge, Governor Frank

Sargent, wonder if he was on the right side of the issue.

In January of 1969, from his corner office, Governor Sargent could see a large crowd of protestors forming in front of the State House. Angry but not violent, they had a very clear message: Don't build the Interstate System through our homes. Most remarkable of all was the solidarity among the diverse group. Irish Catholics, Italian Americans, blacks, Hispanics, and Asians created a sea of voters who represented practically every important part of the Governor's voting bloc.

An hour into the show, a tall, tan, and handsome Governor Sargent emerged from the State House and marched into the crowd.

**In 1969,** Massachusetts governor Sargent promised, "Never, never will this administration make decisions that place people below concrete." He eventually stopped I-95's construction.

He was met with some boos and cackles; but as soon as he began to speak, the protestors listened. He explained that he had just called John Volpe, Secretary of Transportation in Washington—a name that evoked additional antihighway boos—and had told the Secretary that the people of Massachusetts were delivering a message that needed to be heard. The skeptical crowd began to warm up to the idea that their governor was making an effort to connect with them. But then, surprising them all and perhaps himself, Governor Sargent said, "Never, never will this administration make decisions that place people below concrete." It was a crowd pleaser, and the sentiments of the mob made a complete turn as they erupted into hoots and hollers.

It was the biggest decision he would ever make as governor. Some epiphany seems to have occurred with Sargent, or perhaps it was just that 1970 was an election year. Whichever the case, he put a temporary hold on the construction of all Interstate System highways through Boston. In 1971, the next year, he made an even more courageous decision. Following the advice of a task force he had appointed to study the issue further, he killed the two projects for good.

A new idea was in the air. Instead of encouraging more and more cars to come into the city, advocates were pushing the idea of mass transportation—buses and trains moving tens of thousands of commuters a day without the polluting effects of automobile traffic. Driven to create a balanced transportation system of mass transit and highways, the governor chose to refuse hundreds of millions of dollars of potential matching Highway Trust Fund money from Washington, D.C.

Sargent had taken a big gamble, hoping to cash in his Interstate System funds for mass-transit projects. And in the long run he would win. With the Federal Aid Highway Act of 1973, the federal government allowed the states to choose between Interstate System highways and mass transit without being

The nation's capital and 45 state capitals are directly connected to the Interstate System. The capitals not connected are Dover, Delaware, Jefferson City, Missouri, Carson City, Nevada, Pierre, South Dakota, and Juneau, Alaska.

*151*

*Above and left:*
**Stopped cold:** In 1971, more than halfway through a fragile wetland north of Boston called Rumney Marsh, I-95's construction was halted and then lay derelict for 30 years, becoming a junkyard with bridges to nowhere.

*Inset, top:*
**Correcting** its own wrong: Damages to Rumney Marsh have been restored with $3.3 million from a new-generation Interstate project called the Big Dig.

penalized. If they chose to turn in their plans for an urban section of the Interstate System and convert them to a rail system, the earmarked highway funds would remain in the Highway Trust Fund, while other monies would be provided for mass transit.

Sargent's daring move was the start of a national trend. All over the country, people were beginning to realize that the Interstate System was not a miracle cure for the country's ills. And that in some cases, it brought its own road-related demons.

## ROAD RAGE

It was no accident that Boston's Interstate and other Interstates across the country were slated to run through the most vulnerable and least-protected neighborhoods in their cities. Some officials denied

*Above:* **A ghost highway** south of Boston, I-95 was stopped in the middle of a meadow and remains abandoned. *Top:* **As a result of the highway revolt,** today's drivers must negotiate a sharp curve and bypass the city along Route 128 (now called I-95) instead of driving straight through when traveling between Maine and Florida.

**Chan Rogers**
(*right*) made his first superhighway trip on a tank along the Autobahn. After World War II he helped build the Central Artery and then worked on replacing it with the Big Dig's superhighway tunnels.

it in public; but in private, when it came time to place these massive highways, the cities' poorest neighborhoods and their neglected parklands were targeted. Everything was done to avoid accelerating the cities' already dwindling tax bases, and to avoid disturbing the powerful, who might be able to summon resistance that would slow down the forward march of construction. Factories, big businesses, and expensive homes were spared whenever possible. But if you didn't have money or political clout, you were a target. The bulldozers were coming.

"When we were looking for the cheapest place to lay down the Interstates, we would go to the city's tax map that plotted out each lot based on its taxable revenue," recalls Chan Rogers. A Department of Public Works engineer in the early 1950s, Rogers remembers the instructions he received to plot the course of an Interstate. "Using different colors, we shaded in the high-rent districts in one color and the industrial areas in another and the low-rent districts in a third color. That third color just popped out at you, showing us the cheapest route to build along. The other two areas were avoided because they yielded more tax revenue to city hall. Back then it was all about getting from point A to point B with the least expensive acceptable alternative. Today, we would never be allowed to simply take the course of least resistance and cost. But back then it was perfectly acceptable. Hell, we would've been criticized for planning a highway any other way."

What it came down to is that the Interstate System was used, and often abused, as a way for the cities to get rid of their least desirable sections, the slums, while at the same time securing the least expensive right of way for their projects. In 1968 a well-written, well-intentioned study called "Freeway in the City" was published by FHWA as a guidebook for highway officials across the nation, a how-to manual on building the Interstate System through a downtown. The book said, "Some internal freeways have been deliberately located through the worst slums to help the city in its program of slum clearance and urban renewal. The federal government has greeted the concept with enthusiasm." These decisions could have tragic consequences, which tore their cities apart. In some cases, the results are still being felt today.

## RACE RIOTS AND THE INTERSTATE

The brutal and deadly race riots of the 1960s have in part been blamed on the construction of Interstate superhighways. For many blacks, the roads were "white men's roads through black men's homes." The Interstates widened and accentuated the split between a suburban white middle class and an urban black working class. They also made living in the walled-off downtown communities much more difficult. Noise, pollution, and blight were exacerbated by the difficulty of passing over or under the huge superstructures. Many of the city residents didn't own cars and were dependent on mass transit service, which was neglected and in decline. Too often their frustrations boiled over into bloodshed.

In the summer of 1965, Watts, a black section in Los Angeles, erupted into violence that resulted in the death of 34 people. In 1967, Detroit suffered an even deadlier revolt that lasted five days. After the National Guard restored order, 43 people were dead, over 1000 injured, and 7000 arrested. Thirteen hundred buildings were destroyed and nearly 3000 businesses looted. Michigan Governor George Romney later told a Senate committee that freeway construction in Detroit was a major cause of that riot. Neither Detroit nor Los Angeles ever fully recovered from the turmoil of those years or the divisions created by the Interstate System.

## "IF WE HAD IT TO DO OVER AGAIN, WE WOULDN'T"

The targeting of the poor continued across the nation. Claiborne Avenue in New Orleans, just a block from the French Quarter, where some of America's greatest jazz has been performed, is still suffering its consequences.

Many of yesterday's jazz legends walked up and down Claiborne Avenue. They lived there or visited friends and family there, in the heart of this culturally rich black neighborhood of New Orleans. Black store owners often lived in apartments above their establishments on Claiborne Avenue. Their front porches, where many a pickup session was held, faced out onto mighty rows of ancient oaks. The trees were four stories high and, according to some, the longest contiguous chain of oaks in the United States. But that was before I-10 was built through, or rather over, Claiborne Avenue in the early 1960s.

The ancient oak trees, unfortunately, were in the path of the Interstate's elevated highway. And the residents of Claiborne Avenue didn't have the influence to fight the destruction of their neighborhood. So the trees were chainsawed and grubbed out of existence and a rich part of American history was gone forever.

After the damage was done, a couple of highway engineers who had worked on the project were asked what they thought of the I-10-elevated highway. One said it was a "rather grotesque structure" and the other said, "If we had it to do over again, we wouldn't."

In a wistful if ultimately futile attempt to restore what was lost, a group of

**Through New Orleans,** I-10 took the course of least resistance, straight through a poor black neighborhood along Claiborne Avenue, once home to the nation's longest row of oak trees, shown here.

# Superhighways on Stilts

**I**T WASN'T ALWAYS NEIGHBORHOODS that stood in the way of the Interstate. Often Mother Nature herself was the problem, as in New Orleans.

New Orleans sits just four feet above sea level. Dig a few feet into the ground, and the murky waters of the Mississippi River quickly rush in to fill the void. Surrounded by water, the city is more island than mainland.

When the French arrived in the area nearly 300 years ago, they immediately began building moats and dikes to protect their outpost in what became an epic battle of man versus water. Nothing was sacred in this battle. When the earth and its mischievous groundwater returned even their buried dead, the citizens of New Orleans were forced to entomb their loved ones in mausoleums above ground.

Digging a secure foundation for a roadway in this swampy area proved to be even more difficult than digging graves. To build a typical roadbed through a swamp is cost prohibitive. Constructing a foundation would require mucking out up to 30 feet of weak silt and soils, then replacing the muck with a higher quality backfill of sand and gravel. Expensive and time consuming. It's easier to build the highway on stilts, as a bridge over the swampy ground.

The weak soil conditions around New Orleans and the Mississippi Delta require that long stretches of the Interstate System run for miles along these stilted highways, which are called viaducts. But that has its own challenges. Getting heavy cranes and pile drivers into the middle of a marsh is the first one.

Pete Mainville, a highway engineer and contractor for over 40 years has seen just about every construction technique, but building in the swamps sticks out above the others: "You got to get your equipment out there somehow, so the guys built temporary roads, service roads. In the '50s and '60s they used to lash thousands of tires together with cable and then float them out into the swamp. After the tires were laid out, they would dump tons and tons of fill over them. It was like a semi-floating causeway that would allow tractor cranes and trucks to get to the work site. It was the only way to get access to the job."

Building the Interstate System through Louisiana's swamps required more than creativity and some old tires. It required spoons, blow counts, and borings to test the ground for its geotechnical qualities. During the investigation of the highway's right-of-way, a steel rod known as a spoon was driven into the earth by a work crew operating a pile driver. A careful count was made as to how many blows from the pile driver it took to drive the spoon into the earth until it reached ground solid enough to support a highway. Each blow of the pile driver's hammer was a count; and at the end, the blow count along with soil samples gave the engineers a clear idea of design requirements.

In the swamps, the blow counts were almost always low as the piles sank quickly into the watery mix. Foundations up to 80 feet deep could be required in order to reach solid ground.

After the blow counts were enumerated and the engineering drawings finalized, work crews began building the highways. Driving as many as a dozen or more piles in a cluster, they began the task of building the foundations of the viaduct. The pile drivers began hammering in the piles, which ran in length from 30 to 80 feet.

Once a cluster of a dozen or so piles was driven completely below the surface of the earth, a massive cubical concrete cap, four feet thick, was placed on top. This cap, resting just below the surface of the ground, was really a platform to hold up the concrete columns, the ones above the ground supporting the Interstate.

One wonders if the motorists whizzing by on these highways on stilts ever think of the engineering feats that make their journey safe.

local artists recently painted pictures of the mighty Claiborne Oaks on the steel columns of I-10. The man-made oaks now line the avenue, a reminder of what once was.

## THE BATTLE OF BALTIMORE

Not all of the less affluent neighborhoods succumbed to the will of the mighty. Many fought back, and many won. Fells Point and Rosemont, two Baltimore communities, were among the victorious.

U.S. Senator Barbara Mikulski remembers well the battle to save the neighborhoods: "I jumped up on a table, and I cried, 'The British couldn't take Fells Point, the termites couldn't take Fells Point, and goddamn if we'll let the State Roads Commission take Fells Point!'"

Mikulski's actions were hardly the behavior you'd expect of a United States Senator, even a senator-in-the-making, but these were extraordinary times and Mikulski was then a fiery community activist.

In 1968, Maryland's State Roads Commission was about to change life in Baltimore as the citizens knew it. The commission had ambitious plans to plow through several neighborhoods, all lower or middle class, with I-95 and then make a massive interchange in the center of town to connect three Interstates. The Interstates would dissect Baltimore, splitting the city from its waterfront and its financial districts from its historical neighborhoods.

As plans were getting underway, Mikulski remembers getting a phone call from a friend with a plea for help. There was a meeting that night in Fells Point, Mikulski's neighborhood, on the east side of Baltimore. Fells Point was a white, predominantly Polish enclave of homes from the 1700s that had fallen on hard times but were on the verge of a comeback. The meeting was to discuss what to do about the 16-lane highway that was slated to run through the community. "They're afraid to fight the bosses," the friend told Mikulski, "but your name is well known. Come and talk to them."

"So," Mikulski remembers, "I went, and of course I stayed and fought."

Soon after, in the auditorium of a school across town in Rosemont, a proud African-American community, highway officials were holding another hearing. Rosemont was in the path of the same I-95 expressway that threatened Fells

**Defiant art protests** the tragic loss of neighborhood and nature. The mighty oaks sacrificed to the Interstate live on in the form of painted highway columns under the belly of I-10 in New Orleans.

*157*

Baltimore—like
Boston, New Orleans,
San Francisco, and
other cities—won some
and lost some in its
fight against the
Interstate System.
"Proudly Polish," the
bombastic Senator
Barbara Mikulski led
a coalition that saved
ethnic Fells Point
from I-70.

*The
highwaymen's
strategy was
to keep the
neighborhoods
segregated in
an attempt to
divide and
conquer and
ram their
road through.*

Point, but the Fells Point people were not invited to this meeting. The highwaymen's strategy was to keep the neighborhoods segregated in an attempt to divide and conquer and ram their road through.

But Mikulski had her own battle plan. As in Boston, disparate groups found common ground as they fought the politicians.

Struggling to keep the highway from dividing their worlds, east met west. Mikulski helped organize and mobilize two busloads of supporters from her neighborhood. In a show of solidarity, the Fells Point people went to the meeting in Rosemont.

Arriving just as things were getting under way, the Fells Point group saw over 400 black friends, neighbors, and family members seated and prepared to fight for their ground. It's hard to say who was more surprised to see the white faces of the visitors, the highway officials running the hearing or the black community, which was ready to testify to the destruction the expressway would inflict.

In a defining moment, one of Mikulski's group stood up and spoke out, galvanizing the two neighborhoods for good.

"My name is Frank Milkowsky," said the tough dockworker and former merchant mariner, "and I fought in World War II to save America. And I fought to save the government of the United States of America. Now, I'm here to join with the black community, and their veterans and their wives, to save the neighborhoods of Baltimore." "There was thunderous applause," Senator Mikulski recalls. "We clapped, we sang, we cheered. The ice was broken. We were together. Mutual need, mutual respect, mutual identification."

In the end, the Battle of Baltimore was won by the neighborhoods. They successfully beat back the aggressive plans to bring the Interstate System into the heart of their town. As one federal highway official put it, "Baltimore is a better city today for not having done what we wanted it to do back then."

## THE LITTLE OLD LADIES IN TENNIS SHOES

The highway planners not only targeted poor or politically unconnected neighborhoods. In a classic example of nature—and highway builders—abhorring a vacuum, any park without a powerful protector could also be targeted. Parks became especially vulnerable as downtowns declined and there was little money or impetus to maintain them.

One of the best examples of this is what happened, or almost happened, to Overton Park in Memphis, Tennessee. If it were not for a group of "little old ladies in tennis shoes," as the press dubbed them, Overton Park would have been lost forever. But the ladies, a group of elderly women who wouldn't take no for an answer, rallied to save their beloved park from the onslaught of the highway builders. In one of the most significant rulings affecting the Interstate System, *Citizens to Preserve Overton Park* v. *Volpe* was decided by the United States Supreme Court, changing the course of I-40 as it was planned to pass through Memphis and many more of the nation's urban cores.

Overton Park is to Memphis as Central Park is to New York City. Created over a hundred years ago and named after one of the city's founding fathers, the 342-acre oasis is one half natural forest and the other half the setting for a museum, a nine-hole public golf course, hiking trails, ball fields, bridle trails, and a zoo. It is the center of the town's recreational life.

On July 30, 1954, Overton's outdoor amphitheater hosted the first paid performance on a major stage of Elvis Presley, the King of Rock and Roll. A year later the state highway department, with the blessings of the Bureau of Public Roads, planned to plow six lanes of superhighway through Overton. That's when the ladies in tennis shoes started running around town.

The women, dismayed at the thought of I-40 passing through the hallowed grounds of one of the city's few parks, began a furious campaign, collecting over 10,000 petitions to save Overton. It was just the start of a 24-year-long battle that at first seemed hopeless. Nearly everyone wanted this highway built, or so it seemed—starting with the mayor, city councillors, downtown businesses, the chamber of commerce, and the state and federal highway departments.

*Left to Right:*
**While walking** with his daughter, Overton Park resident Bill Bearden often passed homes demolished for a never-built section of I-40. They saw the stairs to nowhere and the abandoned but still blooming flower beds, calling them "haunted spaces." A park saved: I-40's dead end at Overton Park.

The original plans dated back to 1955 and the Yellow Book. They showed that I-40—"America's Highway," a 2554-mile-long Interstate—would wipe out 26 acres of Overton's forest, splitting the park in two with a partially raised and partially depressed highway that spread to a 450-foot width at some points. Ground that was not paved would be destroyed by the noise and by the exhaust from tens of thousands of trucks and cars each day as they passed between California and North Carolina.

In a long line of bureaucratic flip-flops, Overton's future was touch and go through the 1950s and 1960s. By 1970, 842 homes, 44 businesses, five churches, and one fire station had been demolished to make room for what everybody thought was a done deal. But the little old ladies wouldn't give up. Finally, the decision as to whether or not to build the highway went before the U.S. Supreme Court in 1971.

It all came down to one small clause that would alter the course of many more Interstate System highways slated to pass through public parklands like Overton: "feasible and prudent." Congress, in 1966, had said that Interstate highways would no longer be able to pass through public park spaces unless there was no feasible and prudent alternative. In Memphis's case, the Supreme Court ruled that there was an alternative route for I-40, a planned beltway. The route was obvious; and no one, not the most loyal supporter of the highway through the park, could deny it.

On January 26, 1981, the Overton Park section of I-40 was deleted from the Interstate System's master plan. Today, I-40 passes around, not through, Overton Park on what was going to be the city's beltway, and the homes that were leveled have been replaced.

Some still argue that Memphis's downtown was damaged by preventing I-40 from bringing traffic into and out of downtown more quickly. Bill Bearden, a resident living near Overton Park feels differently. He says, "The assassination

The Interstate System is designed to handle cars and trucks speeding along at 50 mph in the urban areas, 60 mph in hilly terrains, and 70 mph in flat rural areas.

160

of Martin Luther King may have killed downtown Memphis, but saving Overton Park by stopping a highway that would've saved commuters a few minutes each day sure as hell didn't ruin it. Besides the downtown is in the midst of a comeback and Overton Park is part of that success." Thanks to those little old ladies in tennis shoes.

## FACING OTHER CHALLENGES

Building through America's cities was one of the biggest challenges facing the forward march of the Interstate system. The times were tumultuous and, when we look back on them now, astonishing in their disregard for the welfare of individual communities in the name of the common good—as defined by the city fathers.

But it was also a time of awakening to the needs of individuals, and communities, the beginning of a monumental civil rights movement. As groups around the nation rallied to protect their neighborhoods and historic places from the bulldozers, the Interstate System echoed this new awareness of the importance of each one of us, and our right to protect our homesteads.

And while the battles for the neighborhoods raged in the cities, across the country another phenomenon was taking place, born also from challenge. As the Interstate crossed the vast space of America, it encountered chasms, rivers, mountains, desert—almost every natural and unnatural landscape one could imagine. The road must go on, but to do so would require nothing less than modern marvels of engineering.

*161*

The San Francisco–Oakland Bay Bridge spans more than eight miles of water, anchoring the western end of I-80.

CHAPTER NINE

# 54,663 Bridges

**B**RIDGES ARE THE SOLUTION to an age-old transportation problem: how to travel over a natural or man-made obstacle. There are 54,663 bridges on the Interstate System, averaging more than one per mile of road. The vast majority of the impediments they pass over are easy to overcome—creeks and streams, other roads, and the occasional livestock crossing. The resulting steel or concrete constructions are short, functional, and nondescript. There are no dramatic approaches, no scenic vistas, no testaments to engineering genius. These bridges are meant to be crossed without notice, save for the annoying bounce of a car's suspension as tires pass over the expansion joints.

But scattered throughout the System are gems, the bridges that have become landmarks in their regions and points of civic pride. The various agencies of the Interstate System recognize these bridges by endowing them with a name—sometimes a very long one—rather than just the standard mile marker and station designation.

Most of these bridges are elegant and beautiful, through natural form if not by deliberate design. A few are rather ordinary, and some are downright ugly to some eyes. Some are visible for miles; they beckon with their long approaches.

Others literally explode on the scene at the end of a tunnel or the other side of a gap. Some offer incomparable views from their approaches or bridge decks, equal to those from America's best scenic roads. Yet unlike the scenic roads, where the panorama may be enjoyed for hours, bridge views can only be savored for fleeting moments at Interstate speeds.

The special bridges highlighted here represent unique answers to difficult engineering challenges—extreme length, long spans, poor anchorage, severe weather. Bridges on the ocean may face strong currents that would sweep away conventional piers. Even inland, many must withstand heavy winds or severe earthquakes. Some of the older bridges predate the Interstate Highway System and its stringent standards. They face problems never envisioned by their creators—higher traffic volumes, heavier and faster vehicles. Old or new, they are all interesting in their way, part of man's never-ending drive to bend nature to his will.

## SUSPENSION BRIDGES, KINGS OF THE CROSSINGS

A suspension bridge is the King of Bridges, capable of spanning the greatest distances between two points. As a result, suspension bridges tend to be the most expensive to build. In recent years, the reign of these kings has been challenged by the modern cable-stayed bridges. But when it comes to spanning great lengths, well over a mile, the suspensions still hold the seat of power.

The modern suspension bridges that we pass over today are supported by large cables made of thousands of steel wires wrapped tightly together. The cables are secured to solid rock or immense concrete-block anchorages on either end, and are draped over two high towers. The bridge deck is suspended from the main cables, using high-strength steel wires. The load of the bridge deck causes the cables to push down on the towers and pull at the anchorages.

Suspension bridges are the lightest, strongest, and most efficient of all bridge types. They are also the most visually pleasing because of their graceful lines and relative simplicity. Their inherent form lends itself to high clearances and long spans, enabling ships to pass safely underneath. The limited number of towers avoids the problem of placing many pier foundations in moving water, where they are subject to undercutting and degradation. So these bridges are ideal for deep-water crossings.

In recent years, suspension bridges have been making a comeback thanks to new developments in material and construction technologies. By late 2003, the new Alfred Zampa Memorial Bridge, carrying I-80 over the Sacramento River north of San Francisco, will become the first new suspension bridge on the Interstate System since 1968.

Nearly 25% of the 54,663 bridges on the Interstate System were built in a five year period between 1965 and 1969.

**The Alfred Zampa**
Memorial Bridge (I-80), the first suspension bridge built on the Interstate System in 35 years. It is named after a colorful local ironworker who helped build many of the original Bay Area bridges, including the span it replaces, the Carquinez Strait Bridge, shown at right.

# THE GEORGE WASHINGTON BRIDGE

**I-95, New York City, New York–Fort Lee, New Jersey, 1931**

When it was built in 1931, this mighty bridge blew the world record for bridge spans out of the water. Regarded by some as the most beautiful bridge in the world, it has twin-deck roadways suspended from four three-foot-diameter main cables atop soaring twin towers. It measures 4760 feet between anchorages, with a main span of 3500 feet.

Known to New Yorkers as the GW, it was designed by the legendary Othmar Ammann, chief engineer of the Port Authority and designer of the Verrazano Narrows, the Bronx–Whitestone, Bayonne, and Throgs Neck bridges. Although the GW's main span was to be twice that of any existing bridge, Ammann's design omitted the stiffening trusses below the roadway that had been considered essential for suspension bridges. Steadfast in his convictions, he felt that as the span increased, the ensuing increased weight of the roadway and cables would be sufficient to resist wind loads. He would be proven right.

After the original six-lane bridge opened to traffic in 1931, two additional lanes followed in 1946. Ammann designed for future expansion, and a six-lane lower deck was added in 1962. At the time, the GW was called the strongest suspension bridge in existence. The same must still be true today, as the traffic volume has increased tenfold in its lifetime, from 5 million vehicles to over 55 million vehicles annually.

The steel latticework of the 604-foot towers is still a sight to be savored. Its intricate, almost lacy design was originally intended to be covered with concrete and granite panels. But public opinion and the economics of the Depression prompted the Bridge Authority to leave it exposed. The GW has also featured the world's largest American flag, hung from the New Jersey tower over the roadways during holidays. It measures 60 feet by 90 feet, weighs 475 pounds, and has five-foot-wide stripes and three-foot-wide stars.

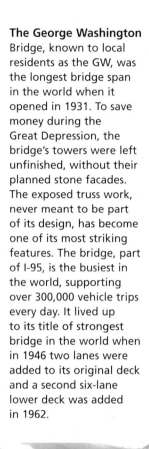

**The George Washington** Bridge, known to local residents as the GW, was the longest bridge span in the world when it opened in 1931. To save money during the Great Depression, the bridge's towers were left unfinished, without their planned stone facades. The exposed truss work, never meant to be part of its design, has become one of its most striking features. The bridge, part of I-95, is the busiest in the world, supporting over 300,000 vehicle trips every day. It lived up to its title of strongest bridge in the world when in 1946 two lanes were added to its original deck and a second six-lane lower deck was added in 1962.

167

# SAN FRANCISCO–OAKLAND BAY BRIDGE
## I-80, San Francisco–Oakland, California, 1936

The "Bay Bridge" is colossal and, along with the George Washington Bridge, one of the busiest in the world when it comes to vehicle traffic. It is really four separate superstructures wrapped into one: three bridges and a tunnel. The 8.4-mile giant spans Oakland Bay, connecting the cities of San Francisco and Oakland. Using the Yerba Buena Island Tunnel as a takeoff and landing platform, the bridge transitions from a cantilever truss bridge on the Oakland side to a stunning double suspension bridge on the San Francisco side of the Bay.

The 9260-foot-long tandem suspension bridges share a common anchorage that acts as a sort of man-made island. The foundations of the center anchorage and four towers were constructed on bedrock in water as deep as 100 feet and through a deep layer of bay mud and clay. The towers rise 526 feet above the Bay, and the 2310-foot main spans provide a vertical clearance of 220 feet above the water.

The 10,176-foot eastern segment between Yerba Buena Island and Oakland is a double-decked cantilever truss bridge with a main span of 1400 feet. It features the world's deepest bridge pier, sunk 242 feet below the water level.

During the 1989 Loma Prieta earthquake, with a magnitude of 7.1, a bolt in the eastern segment failed, causing a 50-foot section of the upper deck to fall onto the one below. The bridge was closed several months for repairs, resulting in massive traffic headaches on both sides of the Bay. After the California Department of Transportation concluded that a seismic upgrade of the existing cantilever truss bridge would be too expensive, it decided to replace the old bridge with the world's first single-tower self-anchored mono-suspension bridge. The reconstruction of the Oakland side of the Bay Bridge heralds a new era, and perhaps the rebirth of the King of Bridges. The new suspension bridge is scheduled to open in 2007.

*Below:* **The Bay Bridge** is one year older and much longer than its prettier little sister, the Golden Gate Bridge, just miles to the west. It is really two suspension bridges, a tunnel, and a steel-truss bridge that create an 8.25-mile crossing.

*Opposite:* **In the 1930s,** an earthmoving rig grades a road on Yerba Buena Island, where a tunnel links the crossing's suspension bridge with its steel-truss bridge.

**When the Mackinac** Bridge was built in 1957, it was overall the longest suspension bridge in the world. Its towers rise 552 feet above the waters joining Lake Michigan and Lake Huron. As part of I-75, the "Mighty Mac" connects Michigan's Upper and Lower peninsulas. The foundations of the bridge were designed to withstand turbulent waters, high winds, and ice buildup.

# MACKINAC BRIDGE

**I-75, Mackinaw City–St. Ignace, Michigan, 1957**

The "Mighty Mac" is the bridge that literally united a state, connecting Michigan's Upper and Lower peninsulas over the turbulent Mackinac Straits where Lakes Superior and Huron meet. Before its opening in November 1957, the only connection between the two peninsulas was an hour-long ferry ride across the straits. In summer, thousands of cars would line up, some waiting an entire day to make the crossing. The Upper Peninsula, much more accessible from the west—from Wisconsin—was for all intents and purposes an extension of that state rather than part of Michigan. The Mighty Mac changed that.

Master bridge designer David Steinman conceived a five-mile-long bridge featuring four travel lanes—two in each direction—atop a deep truss. At the time it was built, the Mighty Mac was the longest suspension bridge in the world, measuring 8614 feet from anchorage to anchor-age. The main 3800-foot span provides a roadway height of 199 feet and allows 155 feet of underclearance at midspan for ships. The towers stand at 552 feet above the water.

When designing the bridge, Steinman also had to consider high wind, heavy snows, the effect of turbulence of the straits on the towers and piers, and pressure due to ice buildup during the cold Great Lakes winters. Wind effects were of particular concern only a decade after the collapse of the Tacoma Narrows Bridge (see page 178). Steinman took several steps to reduce wind-induced oscillation, or galloping. He added a deep stiffening truss beneath the roadway to reduce flexibility. He also provided space between the truss and the roadway and an open-grid roadway beneath the two inside lanes to change the aerodynamic characteristics of the structure.

# CANTILEVER TRUSS BRIDGES: WORKHORSES OF THE INTERSTATE

In science it is said that for every action there is a reaction. This is especially true with the cantilever truss girders, a less expensive workhorse of a bridge.

The longer the span, the deeper its girders must be. But the more steel that is used, the more self-weight must be supported. Therefore cantilever truss bridges use a complex framework of steel girders, enabling them to span long distances. This, however, gives a hulking industrial look to them—pleasing to some but raw and overbearing to most.

Cantilever trusses are a variant of the basic truss. The simplest cantilever truss comprises three components—two outer trusses and a center truss. The outer trusses, which are built first, are anchored at the end abutments and extend over the interior piers, jutting out into the main span. The center truss is then supported by the free ends of the outer trusses. One of the most difficult phases of construction is placing the center truss. On water crossings, the truss is prefabricated on land, shipped to the site on a barge, and lifted into place.

Though not the most attractive of bridge types, cantilever truss bridges provide a useful and economical solution for spans up to 1800 feet.

# RICHMOND–SAN RAFAEL BRIDGE
### I-580, Richmond–San Rafael, California, 1956

The Richmond–San Rafael Bridge was born after the local citizenry got fed up with ferry strikes disrupting transportation across northern San Francisco Bay. On September 1, 1956, the 5.5-mile cantilever truss bridge linking Marin County and the East Bay opened, and the Richmond–San Rafael Ferry Co. went out of business shortly thereafter. When the bridge was completed, it was one of the largest in the world.

The twin-deck bridge features two 1070-foot spans crossing the main and secondary shipping channels, providing 185-foot and 135-foot clearances respectively. A brute of a bridge, it has been struck several times by passing ships—including twice in a single day—without sustaining major damage.

Compared to the landmark bridges of the Bay, the Richmond–San Rafael Bridge is less elegant and more functional. With its distinctive dip between the main spans, it has been likened to a bent coat hanger and dubbed the "roller-coaster bridge." Even the Marin side approach is harsh, passing by the notorious San Quentin Prison. For all of its lack of appeal, the bridge is a vital link; and for vacationers coming from the northeast on I-80 or from the south on I-80 and I-580, it is the gateway to Marin County.

Although it's officially named the John F. McCarthy Memorial Bridge, very few Bay Area residents would recognize the name, preferring to call it by the names of the two cities that it joins. A drive across it at sunrise and sunset is unmatched as the still water ignites in orange and the multitude of girders creates a dizzying display of angular shadows.

**The Richmond–San Rafael Bridge** is the portal to the East Bay and the Interstate's link to the rest of the nation. Despite its 185-foot-high clearance, it has been hit by many ships (it was "bumped" by two naval vessels in one day) but has never been closed as a result.

# ARCH BRIDGES

Arch bridges are among the oldest of all bridge types. Arches use a curved beam, which is fixed horizontally by bearings, or abutments, on both ends. Vertical loads—in this case, the weight of traffic—are transferred through the arch and carried outward toward the abutments. The abutments resist the tendency of the arch to spread apart at the ends and sink at midspan. Because of this, arch bridges can only be used in locations where the bearings rest on solid or stable ground.

Arch bridges may have two or three hinges that allow rotation. Two-hinged arches, which have a hinge on each abutment, are very economical and the most common type used for steel bridges. The three-hinged variety adds another hinge to the top of the arch, and is less vulnerable to the effects of a shifting foundation due to earthquakes or ground subsidence. However, they are harder to construct and are rarely used.

Another variation is the tied arch. Instead of using abutments to resist the horizontal forces and the tendency of the arch to spread, a tied arch uses a girder to tie the ends together. The girder acts in a manner similar to the bowstring on an archery bow. Tied arches are ideal for locations where the ground is not solid or stable.

The depth of the arch allows for spans of up to 1500 feet, but an additional appeal of this bridge type lies in its aesthetics. The curve of the arch and the overall simplicity of design make for a bridge of great beauty and elegance, while its vertical nature is ideal for use in deep ravines or gorges.

# FREMONT BRIDGE
## I-405, Portland, Oregon, 1973

The double-decked Fremont Bridge is the only major structure on I-405, a four-mile-long offshoot of I-5 that loops around downtown Portland on the west side of the Willamette River. At 902 feet, the bridge is the longest tied-arch bridge in the world and has the longest main span of any bridge in Oregon. Its novel three-span tied arch is a solution to site conditions; but it conveys an image of gracefulness and lightness that makes the bridge stand out on the skyline.

To minimize disruption of river traffic and lower the cost, the center span was built off-site and floated on the river to the bridge site. There it was lifted into place from the river below using 32 hydraulic jacks. The 6000 tons lifted set a world's record for the heaviest lift.

# GIRDER BRIDGES: THE ORIGINAL

Girder, or beam, bridges are the simplest and most common type of bridge on the Interstate System. They are probably the oldest kind of bridge in the world, dating back to the first time someone gingerly stepped onto a felled tree to cross a stream. The simplest girder bridge is a bridge deck resting on girders supported on each end by abutments.

The girders vary by cross section and materials. The most common cross sections are the I-beam, the T-beam, and the box girder. While I-beams and T-beams make efficient use of material, box girders are more stable and resist twist. Box girders are generally used for bridges that are curved. Beams and girders traditionally have been constructed with steel or concrete, but new materials are on the horizon.

Girder bridges are used on short spans—ranging from 30 to 600 feet between supports. With longer spans, the depth of the girder increases. To keep them from being too deep, girders may rest on one or more piers or trestles between the abutments. This bridge type is known as a multispan continuous-span bridge or viaduct. On Interstate highways, the simplicity of steel or concrete box-girder viaducts may be seen at its best at multilevel interchanges and on approaches to large bridges.

# BERNARD F. DICKMAN BRIDGE
## I-70, St. Louis, Missouri, 1967

The Dickman Bridge, a.k.a. the PSB—an acronym for the Poplar Street Bridge—is the major Mississippi River crossing in St. Louis, carrying I-70, I-55 and U.S. 40. The bridge is 2165 feet long and is supported by abutments and four piers. Its main structure consists of four box-girder sections of varying depths ranging from 10 to 12 feet at midspan to 20 to 25 feet near the piers.

At first glance, the PSB appears typical of its genre—a simple design and a true workhorse. However, this bridge has several features that make it unique. It is one of the few steel box-girder bridges in the United States with an extremely thin and durable deck of continuous orthotropic steel. The deck, which integrates the driving surface as part of the structure, is only 9/16 of an inch thick and very light. The PSB also features a gondola car, suspended on a monorail beneath the deck, that is used to inspect the outside of the girders and to collect river data for the U.S. Geological Survey.

While the bridge itself isn't striking aesthetically, it provides nice evening views of St. Louis, the Gateway Arch, and Busch Stadium for westbound travelers.

If not famous for its beauty, the PSB is at least notorious to local commuters for its congestion. Located near the intersections of four major interstates (I-70, U.S. 40, I-55 and I-64), its eight lanes convey over 125,000 vehicles daily. A new 2000-foot-span cable-stayed bridge has been proposed to be built north of the PSB to alleviate the traffic burden, with a projected opening in 2010.

## H-3 VIADUCTS

### H-3, North Halawa Valley–Hospital Rock, Hawaii, 1996

With a price tag of 1.3 billion dollars, the 16.1-mile Interstate H-3 was arguably the most expensive road in America in terms of per-mile cost when it was completed in 1996. The term "interstate" may appear to be a misnomer, since the road never leaves the island of Oahu, much less the borders of the state. However, the original interstate highway system is actually a system of both interstate roads and national-defense access highways. H-3 fulfills the latter requirement, connecting the Pearl Harbor Naval Base on the leeward (south and west) side of Oahu to the Kaneohe Marine Corps Naval Air Station on the windward (east) side. Hawaii has a unique system for numbering such routes, using an H- designation rather than the usual I- with the route number.

H-3 is one of the most spectacular highways in America. The roadway glides over the North Halawa Valley, through the Koolau Range via the mile-long Trans-Koolau twin tunnels, into the lush farmlands of the Haiku Valley, before ending at the main gate of the Marine Corps station. The true showcases of the highway are the North Halawa Valley Viaduct and the Windward Viaduct located on the leeward and windward sides of the Trans-Koolau tunnel respectively.

The North Halawa Valley Viaduct consists of two separate parallel bridges extending from the North Halawa Valley roadway to the tunnel entrance. The outbound bridge is 5400 feet long, and the inbound side is 6230 feet long. They are supported by box piers whose heights range from 27 to 105 feet depending on the local topography. The spans are of varying lengths, with the

longest 360 feet. The bridges of the Windward Viaduct are each 6600 feet in length with a maximum span of 300 feet.

Both viaducts were built to have the least impact on the archaeologically and historically rich terrain beneath them. The piers of the leeward viaduct were located between two adjacent sites considered sacred by some Hawaiians. On the windward side, the concrete used was colored so that it would blend into the surrounding Koolau range. The result is that the two graceful roadways soaring above the rugged terrain are environmentally and culturally sensitive, seismically strong, and provide unparalleled views of the beautiful Hawaiian landscape.

*Opposite:* **Avoiding sacred burial grounds,** the bridge piers of H-3 touch the earth of the North Halawa Valley as little as possible.
*Above:* **Hawaii labels** its interstates H-1, H-2, H-3.

# Bridge Disasters

**W**ITH OVER 54,000 bridges on the Interstate System, something is bound to go wrong with some of them sometimes. And when a bridge fails, it can fail in a big way.

## TACOMA NARROWS BRIDGE

Sixteen years before work began on the Interstate System, a single event had a dramatic impact on what would be the network's greatest superstructures, its suspension bridges. On July 1, 1940, the newly built Tacoma Narrows Bridge opened as the pride of Puget Sound. Its 2800-foot span was the third longest in the world. It was known throughout the engineering community for its graceful lines and minimalist design. When locals began driving across it, they noticed a rolling motion in the deck. Their vehicles seemed to be climbing and descending hills as they traveled over the bridge's deck. Finding the quirk endearing, not concerning, residents nicknamed the bridge Galloping Gertie, helping to make it an instant tourist attraction.

The fun turned tragic on November 7, 1940. A strong 44-mile-per-hour wind blew down the Narrows of Puget Sound and against the side of the bridge's deck. What was at first a gentle longitudinal undulation became a transverse twisting movement; or, in lay terms, the bridge's deck first experienced a gentle sway, which quickly turned into a violent rocking.

Urgently, the bridge was closed to traffic; but not before a car carrying a newspaper reporter and his dog and a truck with two people in it were trapped on the deck of the bridge, which was now twisting so much its surface reached 45-degree angles to the waters below.

The newspaperman's car slid sideways across the bridge slamming broadside against the curb. The impact threw open his door, dumping him on the sidewalk. Lucky to have survived, he picked himself up and scrambled off the bridge, leaving his dog behind. After other turbulent undulations knocked the truck onto its side, its occupants also climbed out of the wreck, making their way off the bridge.

A famous newsreel captured the men's plight on the bridge. The only escape route was down the centerline of the deck, which was the part of the road's surface least affected by the violent swinging. In the end, gunshotlike blasts rang out as the thick suspension cables began to snap, sending the bridge's street lamps crashing down on the deck. Within minutes the deck tore itself apart, a 600-foot section of its two-lane road plunging into the waters of the narrows. Remarkably, the only victim was the unlucky canine in his master's abandoned car.

The bridge was just four months old when it

**Tacoma Narrows Bridge** filmed during the oscillation that led to its eventual collapse.

**Tacoma Narrows** after the collapse of the roadbed.

went from famous to infamous. Every suspension bridge to be built, and some of those already built, was impacted by the disaster. The bridge's most appealing feature, its thin main span, was its fatal flaw. The deck was so thin and light—a trend in suspension-bridge design at the time—that the wind was able to blow the deck into a swinging motion. On that fateful day, persistent winds started a sway that, once started, wouldn't stop. A heavier deck might not have ever started swaying; and even if it had, it would have arrested itself as soon as the wind died down.

The bridge was salvaged and its deck was

replaced with a stiffer, heavier, and deeper main span. Its aerodynamics were also improved, allowing the winds to pass through it and not up against it, reducing the reckless swaying.

The disaster sent shock waves through the bridge-design community, and existing suspension bridges were reexamined and in some cases retrofits were made. The design of the next great suspension bridges to be built, the Mackinac and the Verrazano Narrows bridges, would be impacted greatly. Today the Federal Highway Administration manages its own wind tunnel at the Turner Fairbank Highway Research Center in McLean, Virginia, as a result of the Tacoma Narrows collapse. In fact, to this day, every suspension bridge on the Interstate System takes into account the disastrous fate of Galloping Gertie.

## SILVER BRIDGE

At rush hour on December 15, 1967, the Silver Bridge, a half-century-old eyebar-chain suspension bridge over the Ohio River, collapsed without warning. Forty-six people died when a single suspender snapped, causing several others to fail, sending the deck of the bridge and all the vehicles on it crashing down onto the river's banks and into its icy waters. Although the bridge was not part of the Interstate System, its demise put into effect a rigid bridge-inspection program that directly impacted the System.

The Silver Bridge's well-used 700-foot main span crossed the Ohio River between Point Pleasant in West Virginia and Kanauga in Ohio. Later, forensic engineering determined that the failure was caused by fatigue at a point of high stress in an eyebar chain. The problem was that the fatigued area could not be seen, as it was buried in the inner workings of the structure. As a result, today all critical structural points on all Interstate System bridges must be easily seen, making for accurate visual inspections.

Disasters like this become a catalyst for change, forcing actions that may not be popular or appear urgent until there is death and destruction. At a time when the Interstate System construction was at its peak, the Silver Bridge disaster radically altered the

**Silver Bridge** after the collapse.

bridge-inspection process. No one argued with the strict new regulations even though the states were forced to inspect every bridge they owned at least once every two years.

Today the inspection protocol is known as the Highway Bridge Replacement and Rehabilitation Program. At the heart of the effort is a complete list of the loading capacity, functional obsolescence, and structural soundness of every bridge in the nation. States are required to submit information about the condition of their bridges and tunnels to the Federal Highway Administration, which manages the program's national data bank. Fortunately, problems in other bridges have been addressed before catastrophic failure, and to this day there has not been another collapse due to fatigue leading to such great loss of life.

## MIANUS RIVER BRIDGE

In the darkness of the early morning on June 28, 1983, Bill Anderson and Shannon Kelly of Atlanta, Georgia, were heading north on I-95 in Connecticut. Just ahead of them was a tractor-trailer truck that suddenly began braking and then jackknifing. As quickly as the brake lights and dark blue smoke of skidding rubber tires appeared, they disappeared into an abyss.

Brakes slamming, the couple's car came to a stop just 10 feet from tragedy. A 100-foot section of the highway's bridge had dropped off, putting two trucks

and two passenger cars into a 75-foot free fall into the Mianus River below. In the aftermath, three people were dead and three more were critically injured.

The Mianus River Bridge was a plate-girder type with a series of spans held together by pins—rusted pins as it turned out. A section of the bridge carrying all three northbound lanes of traffic failed when the decayed pins holding it in place broke apart without much warning, although residents living near the bridge claimed they heard weird groaning noises long before its deadly collapse. Regardless, people were dead or seriously injured; and an entire summer of traffic jams, miles and miles long, was in store for legions of vacationers heading to the New England coast.

As with the other bridge failures, the Mianus River collapse led to much soul-searching and reevaluation among bridge engineers. The chief lesson from this tragedy was that bridge designs have to be redundant. If one main load-carrying system fails, there needs to be a second system in place.

As important as a call for higher standards in construction was a call for a stricter inspection process. Claims were made that the inspectors in charge of the Mianus River never got out of their cars to take a close look at the bridges they were responsible for. As a result, bridges of this and similar designs now have catwalks built into them so inspectors can get close up to uncover threatening conditions.

## SCHOHARIE CREEK BRIDGE

New York State's Schoharie Creek is one mean and dangerous body of water. In spite of its normally innocent and calm appearance, it's capable of becoming a torrent of death and destruction when heavy rains fill its banks with angry brown water.

In April of 1987, after days of rain, a local sheriff did the only wise thing for a public safety officer to do. He shut down all the bridges running over the creek that were under his control. Unfortunately, the I-90 Schoharie Creek Bridge on the New York Thruway wasn't under his jurisdiction, though it passed through his county, and it remained open to traffic. On April 5, ten people were killed when the foundations of the bridge were swept out from under the superhighway by powerful floodwaters. Enormous sections of the destroyed superstructure were propelled 80 feet downstream.

The bridge's design and slack inspections are said to have triggered the disaster. Locals believe that had Interstate System engineers taken local history more seriously, they would have placed the bridge's foundations anywhere but where they did—in the middle of the creek.

The Schoharie Creek Bridge wasn't the first bridge to be swept away by floods. Contributing to the catastrophe was a breakdown in the inspection process, leaving a serious flaw undetected. Thirty-five years of exposure to rushing water had eroded and weakened the bridge column's foundations. But according to one account, the closest bridge inspectors got to inspecting the vital supports was a trip across the bridge's catwalks, 70 feet above the water.

**I-40 bridge** after the collapse.

Once again, the collapse of a bridge greatly impacted the way bridge designers and inspectors went about their work. This time the focus was on the Interstate System's bridge piers and abutments. Inspectors were required to get wet and get close, employing diving techniques to determine the condition of footings. Formulas were developed to predict the damage around footings for existing and new bridges. Instruments were developed to measure the damage in place. As with other failures, the improvements were codified and incorporated for all bridges built as part of the Interstate System.

## OKLAHOMA'S I-40 ARKANSAS RIVER BRIDGE

Sometimes nothing within reason can be done to prevent a tragedy. At 5:30 a.m. on Sunday morning, May 26, 2002, Joe Dedman, a 61-year-old barge master, began his day just as he had for the previous 30 years, navigating a river barge. This morning's task required maneuvering up the Arkansas River with two empty barges lashed together side by side. Just over two hours into his daily routine, Captain Dedman, in apparent good health, did something he had never done before. He passed out.

His blackout lasted for only a couple of minutes, but that was enough to prove fatal. Veering 300 feet outside of the marked channel at five miles an hour, the barge rammed into the supports of a nameless bridge over I-40. With the force of 62 one-ton cars simultaneously slamming into the gigantic bridge column, the barge severed the bridge's foundation, sending all four lanes of America's busiest east–west trucking Interstate highway into the river. When the captain came to, he was in a living nightmare. A section of I-40 was resting on his empty barges as tractor-trailers, cars, and trucks were still plummeting from the highway more than six stories above. Several people were rescued, but it would be several days before all 14 victims could be recovered from their vehicles.

The bridge, built in 1967, was in "great shape" according to its owner, the state of Oklahoma. The record showed that the structure had been inspected just the year before. The tragedy was simply a freak accident. No overhauls of the bridge-inspection system were called for, and the bridge design was not the point of controversy. Instead, the tragic collapse set a standard for speed and efficiency in reconstruction. In just two months—ten days ahead of schedule—traffic was flowing along I-40 again. Mary Peters, the head of the Federal Highway Administration, announced, "Most magicians are content to pull a rabbit out of a hat. You folks in Oklahoma have pulled a bridge out of a river in record time."

## CABLE-STAYED BRIDGES: NEW BUT OLD

With their clean lines and sharp angles, cable-stayed bridges look like modern sculptures, but their lineage can be traced back to the early seventeenth century, when a Venetian carpenter built a timber bridge with chain stays. This type of bridge became popular in Europe after World War II when the Germans and other nations set out to rebuild their thousands of bombed bridges with money supplied by the United States and its Marshall Plan.

The first cable-stayed bridge in the United States, the Sitka Harbor Bridge in Alaska, was built in 1975 and was soon followed by other examples. Cable-stayed bridges are beginning to supplant other bridge types in medium and long spans between 500 and 2800 feet.

Cable-stayed bridges and suspension bridges both feature towers and suspended roadways. There the similarity ends. A suspension bridge's massive suspension cables drape over two towers, and hangers connect the road deck to the suspension cables.

On a cable-stayed bridge, the large suspension cables and many cable hangers are replaced with strong cable stays. Running directly between the bridge tower and the bridge deck, these cable stays are many wires wrapped together to make strands, and multiple strands make a single cable stay. The cable-stayed bridge is clean in design with a simple angular appearance.

Cable-stayed bridges have become a favorite among state highway departments because they are visually pleasing but cost effective. They use less cable than suspension bridges, don't require expensive anchorages, and are much faster to build. For that reason, more and more are popping up along the Interstate System, often replacing older bridges.

*182*

## HALE BOGGS MEMORIAL BRIDGE

**I-310, Luling–Destrehan, Louisiana, 1983**

Built in 1983, the Hale Boggs Memorial Bridge claims many firsts. It is the first high-level, long-span cable-stayed bridge built in the United States; the first cable-stayed bridge on the Interstate System; and the first cable-stayed bridge to cross the Mississippi River. The bridge features cables radiating in a fan arrangement from the top of two A-shaped pylons, to support a main span of 1220 feet. It was designed to withstand

**The Hale Boggs** Memorial Bridge was the first cable-stayed bridge on the Interstate System and the first such bridge to cross over the Mississippi River. Its success encouraged the use of cable-stayed bridges all over the United States.

hurricane-force winds up to 200 miles per hour that might blow off the Gulf of Mexico.

The bridge connects the north and south banks of St. Charles Parish outside of New Orleans, and has helped open the southern area of the parish to commercial and residential development. It replaced a state ferry, the George Prince, which was upended by a tanker in 1976, killing 77 people.

It is only fitting that the first cable-stayed bridge on the Interstate System should be named after Louisiana native Representative T. Hale Boggs. In the House, Boggs spearheaded the efforts to create a funding mechanism for the Interstate Highway System. In 1996, on the fortieth anniversary of the system, Congressman Boggs was posthumously honored as one of four Visionaries of the Interstate, along with President Eisenhower, Senator Albert Gore, Sr., and Federal Highway Administration head Frank Turner.

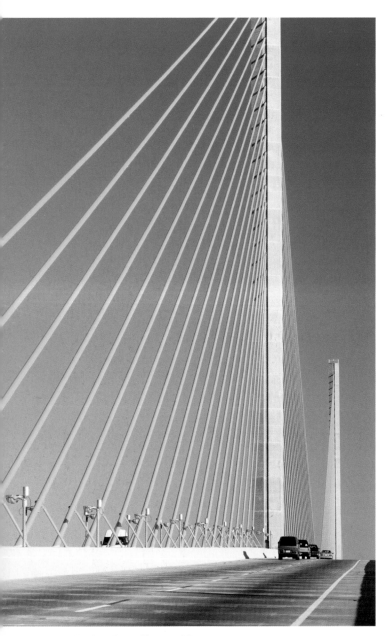

**The bright yellow cables** of the Sunshine Skyway Bridge have become a Florida landmark. Sometimes referred to as the most protected bridge in the world, it's built to withstand hurricane-force winds, and its concrete "dolphins" are designed to fend off 87,000-ton ships.

## SUNSHINE SKYWAY BRIDGE
### I-275, Saint Petersburg–Bradenton, Florida, 1987

At 29,040 feet in length, the Sunshine Skyway Bridge is the longest cable-stayed concrete bridge in the world. Twenty-one bright yellow cables radiating in a fan arrangement from single pylons support roadways on either side of the cables. It offers the motorist an amazing view on its approach, as the road appears to take off like an enormous ski jump and merge with the pylons.

The Sunshine Skyway Bridge may be best known for the twin-span steel cantilever truss bridge that it replaced. During a violent storm in May 1980, a freighter struck the No. 2 South Pier near the center of the southbound span. The collision knocked out 1000 feet of the main span and approach into Tampa Bay, along with several cars and a Greyhound bus. Thirty-five people, most of whom were on the bus, died in the accident.

The cable-stayed replacement bridge was opened in 1987. The new design has a main span of 1200 feet and a 190-foot clearance above the water at midspan. A model of it was tested in a wind tunnel to confirm its safety in hurricane-force winds, and it features large concrete islands known as dolphins to protect the piers from collisions. The dolphins were engineered to absorb the impact of an 87,000-ton vessel and, its designers hope, will prevent a recurrence of the 1980 disaster.

Unfortunately, the new Sunshine Skyway Bridge has not fared much better than its predecessor. The romantic image of the phoenix rising out of the ashes has been tarnished as the bridge operators have spent millions of dollars to fix corrosion problems discovered 14 years into an intended 75- to 100-year life span.

# LEONARD P. ZAKIM
# BUNKER HILL BRIDGE

**I-93, Boston, Massachusetts, 2003**

Of all the bridges on the Interstate System, the brand new Leonard P. Zakim Bunker Hill Bridge is last but not least the 54,663rd bridge on the System. Crossing the Charles River at the northern end of Boston's Big Dig project, it is the elegant aboveground signature of the most ambitious construction project in the history of the Interstate System.

At 183 feet wide, it is the widest cable-stayed bridge in the world. It carries ten lanes of I-93 in an asymmetric arrangement—eight lanes under the inverted-Y pylons, two lanes off the east side. It is the world's first hybrid cable-stayed bridge, with the 745-foot main span consisting of a steel box girder and steel floor beams, and the back spans containing post-tensioned concrete. Its 116 main cables radiate from the 300-foot-plus high towers in a fan arrangement. The main span is supported by two planes of cables from each tower; the back spans are each supported by a single plane of cable stays.

Approached on I-93 from the north, the Zakim Bridge is an unmistakable part of the skyline, with the tops of its pylons a tribute to the nearby Bunker Hill Monument. At night the bridge is illuminated with breathtaking effect. From the south, it literally bursts into view at the end of the new northbound I-93 tunnel. Its soft blue lights and the sharp lines of its cables and concrete obelisk towers create a refreshing contrast to the surrounding historic redbrick structures for which the city is famous.

**Called the Big Dig's** crown jewel, this bridge caps off the most complex and expensive project on the Interstate System. One of the world's most outstanding bridges, it is the widest cable-stayed bridge and the first to have an asymmetrical deck. The old green truss bridge to the right will be torn down when all 10 lanes of the new span are open. The bridge's namesake, the Bunker Hill Monument, can be seen off to the far right.

# UNIQUE BRIDGES

Sometimes the best solution to a difficult engineering problem results in a unique bridge. These designs may be driven by a number of factors: long spans, heavy traffic loads, unusual site conditions, and, of course, economics. The following bridges each had a unique problem to solve. Both solved their dilemmas with novel approaches; the success of the concepts is reflected in the impact that each bridge has had on its respective community.

# LAKE WASHINGTON FLOATING BRIDGES
**I-90, Seattle–Mercer Island, Washington**
**Lacey V. Murrow Memorial Bridge, 1940/1993**
**Homer M. Hadley Memorial Bridge, 1989**

Yes, they really are floating!

The 6620-foot twin floating bridges connect Seattle's western terminus of I-90 and Mercer Island on Lake Washington. From Mercer Island, I-90 continues to the suburbs of Renton, Bellevue, and Issaquah and beyond. The original floating bridge was opened in 1940 and named after Lacey Murrow, the former head of the Washington State Department of Highways. Its four-lane roadway was supported by 22 concrete pontoons, each 14 feet deep, secured by cables connected to the bottom of the lake.

Designer Homer Hadley advocated the use of a floating bridge because Lake Washington is too deep to hold any supporting structures. Reaching depths of 200 feet of water along with another 200 feet of soft muck before reaching stable ground, the lake made building a traditional foundation impractical. As an unintended side benefit, the water protects the concrete pontoons from the ground during earthquakes.

When it was built, the bridge had a set of reversible lanes for peak travel periods and a drawbridge section. Its opening spurred the development of the east side of the lake; and by the 1980s, heavy traffic volume led to the construction of a second floating bridge to the north of the original. The newer Homer M. Hadley Memorial Bridge carries the westbound lanes of I-90, two reversible lanes, and a bike path.

A year later, a section of the Murrow floating bridge collapsed when a pontoon filled with water during a severe storm. The pontoon sank to the bottom of the lake pulling a 2800-foot section of the old bridge down with it. Fortunately the bridge was closed for maintenance and no one was killed. A restored Lacey V. Murrow Memorial Bridge, using replacement sections built with high-performance concrete, was opened in 1993.

Floating-bridge advocate Homer Hadley was a brilliant engineer and visionary. Besides the Murrow Bridge, Hadley designed one of the first paving machines in the United States, the first concrete box-girder bridge in the United States, and several buildings in Alaska that withstood the 1964 Anchorage earthquake, not to mention a number of steel bridges.

**The Homer Hadley Bridge** floats over Lake Washington. The rebuilt Lacey V. Murrow to its right was converted to a one-directional bridge after its pontoons were swamped in a storm and a large section sank.

## MONITOR AND MERRIMAC MEMORIAL BRIDGE-TUNNEL

**I-664, Newport News–Portsmouth, Virginia, 1992**

This is a bridge that turns into a tunnel to allow the navy's largest ships—coming and going out of its nearby shipyard—to head out to sea without interference from a bridge.

The superstructure is named after the two famous Civil War ironclad ships that battled in Hampton Roads Harbor. History purists claim that its name is a grievous misnomer. The battle was between the U.S.S. *Monitor* and the C.S.S *Virginia,* which was built using the scuttled hull of the U.S.S. *Merrimack.* Regardless, the two vessels sank in nearby waters after inflicting mortal wounds on one another.

The 4.6-mile bridge-and-tunnel complex uses a novel approach to allow I-664 to cross Hampton Roads without hindering water traffic. From Portsmouth, the four lanes of I-664 cross the waterway on twin 3.2-mile trestle bridges before diving into a 4800-foot long tunnel that takes them safely beneath the shipping channel just south of Newport News.

This bridge-tunnel was built to alleviate the heavy traffic on the nearby Hampton Roads Bridge-Tunnel. When the new road opened in 1992, the volume was less than anticipated; but drivers soon discovered it, and traffic has grown steadily since. By 2010 it could approach the M&M's design capacity of 75,000 vehicles a day.

## CHALLENGES AHEAD

Of the bridges highlighted here, several are in the process of being replaced or augmented, and one is being considered for expansion. Some have had emergency repairs performed on them in the past several years because of structural deficiencies or outright failures. These problems are not limited to the older bridges, as the example of the Sunshine Skyway so clearly illustrates. And if corroded cables, cracked welds, failed pins, scoured foundations, metal fatigue, and ship collisions weren't enough, the events of September 11, 2001, added terrorism to the list of potential bridge threats.

Other bridges are functionally obsolete, victims of population explosions and the ensuing increase in traffic volume. Marvels of engineering and aesthetics have been transformed into mere bottlenecks in the eyes of angry commuters.

The challenges facing these bridges are no different from those confronting all 54,663 bridges on the Interstate System. The General Accounting Office recently conducted a survey of highway officials in all 50 states, the District of Columbia, and Puerto Rico on the role and condition of the Interstate Highway System. Its report revealed that half of the Interstate bridges are over 33 years old and many are nearing the end of their life spans.

But there is cause for hope. The same GAO report indicated that the number of structurally deficient Interstate bridges decreased by 22 percent from 1992 to 2000. Engineers have a greater understanding of failure and aging mechanisms. Bridge agencies are improving

**Drivers pass over or under** most of the Interstate System's 54,663 bridges without a thought. I-25 hosts a four level interchange south of Denver.

the inspection process and are using remote sensors to monitor load levels and detect corrosion and flaws. The technology of repair is improving with the introduction of better techniques, materials, and coatings.

The outlook for new bridges is bright as well. Bridges of the future will be constructed of advanced materials—composites now used on aircraft and cars, high-performance concrete, fiber-reinforced polymers to name a few. These high-strength, lightweight, and corrosion-resistant materials have breathed new life into old forms, as exemplified by the self-anchored suspension segment of the new Bay Bridge. The interstate bridges of tomorrow—be they unnamed girder bridges or classic landmarks—will solve their engineering challenges with beauty, strength, and reliability.

In 1985 the construction of the Fort McHenry Tunnel was the largest single project on the Interstate System, and cost $750 million.

CHAPTER TEN

# 104 Tunnels

TUNNELS ARE AS SCARCE on the Interstate System as bridges are numerous. There are only 104 tunnels in the system, compared to the 50,000-plus bridges. There's a simple explanation for this. Tunnels are expensive, very expensive. A tunnel can easily cost three times as much as a bridge of equal capacity. Despite their cost, these underground dark holes, disappearing into mountains or hiding under waterways, are highly desirable pieces of construction.

For reducing hazards to navigation for ships and aircraft, nothing beats a tunnel. When trying to reduce the effect of traffic—pollution, noise, ugliness—a tunnel is your best bet. If tunnels were not so expensive, they would be everywhere on the Interstates. In the world of infrastructure, a tunnel is often a sign of a city's transportation sophistication.

The key to these underground wonders is ventilation. Tunnels have been around ever since man first learned to burrow through rather than around an obstacle. But the mechanics of bringing fresh air into a tunnel and removing the old air is a modern requirement made necessary by our desire to run combustion engines through them. Only in the last 75 years has man begun to understand how to create an inhabitable underground environment.

The Interstate System has led the world into the next generation of tunnel building. It has become a showcase for these modern marvels.

## THE HOLLAND TUNNEL

**The Holland Tunnel,** the first underwater tunnel for automobiles, took seven years of backbreaking work and cost 13 workers their lives. Here, two sandhogs tighten bolts, securing one of the iron rings that make up the tunnel's lining. On a good day, the men could place 40 feet of new iron walls.

In 1906, New Jersey and New York decided to eliminate the gap between them by building a bridge over the Hudson River. At the time, it was impossible to drive a car from New Jersey to Manhattan Island. There wasn't a single bridge or tunnel for automobile traffic. A motorist heading into the city from New Jersey had one choice, the Hudson River ferry service.

What was conceived as the first bridge over the Hudson River between New York City and New Jersey turned out to be the first underwater tunnel in the world built for automobiles. Today, depending on how you look at it, that tunnel is the beginning or end point of I-78, and it holds the claim of being the oldest piece of major infrastructure on the Interstate System. The world knows it as the Holland Tunnel, the longest and widest underwater tunnel of its time.

The Holland Tunnel was also the first tunnel that could boast a mechanical ventilation system. Clifford M. Holland, the young man whose name is mounted on the entrance of the tunnel, was the engineering genius who made the historic path beneath the Hudson River a reality. Holland's original ventilation system is still in use there, and nearly every tunnel built on the Interstate System continues to employ his theories and practices.

After seven years of debates about the location of a bridge, and with concerns about its being a hazard to the navigation of ships as well as its vulnerability to inclement weather, state officials decided to cross beneath the Hudson rather than over it. In 1919 they turned to Holland to help them do it.

Holland had already built subway tunnels under New York's East River, but this was different. Unlike the clean electrical power and thin bodies of the subway trains in his former tunnels, the hefty, emission-spewing motorized vehicles demanded more room and a system to take lethal exhaust fumes out of the tunnel's confined space and bring fresh air in.

To get a better idea of just what was required of a ventilation system, Holland worked with the United States Bureau of Mines, conducting over 2000 tests exploring the human body's tolerance to carbon monoxide. Working backward from there, Holland's teams calculated that the tunnel would need to replace all of its air every 90 minutes. It was a daunting job, and many engineers thought he was crazy to even try.

## BREAKTHROUGH

But he did it. To this day Holland's original 84 fans replace all of the air in the mile-and-a-half-long tunnel every 90 minutes, as was deemed necessary. The breakthrough in his design was a three-tiered tunnel: one large chamber for traffic and two smaller ones above and below it. The bottom tier, called the supply-air plenum, carries fresh air into the tunnel system, powered by 42 of Holland's fans. After the fresh air mixes with the vehicles' carbon monoxide emissions, it is sucked out of the tunnel through vents in the ceiling and into the top tier, called the exhaust plenum, from which the other old 42 more gigantic fans pull the air out of the tunnel.

Sadly, Holland died before his tunnel was completed. At the age of 41, he had worked himself to death. In honor of his contributions, the tunnel to which he gave so much carries his name.

On November 12, 1927, President Calvin Coolidge, while floating on his yacht on the Potomac River, pressed a button that triggered the ringing of a large brass bell at the entrance of Holland's tunnel, signaling its opening. That night a truck making a delivery to Bloomingdale's, the Manhattan department store, was the first vehicle to pass under the waters of the Hudson River through the tunnel. Holland's engineering marvel was added to the Interstate System in 1955, with its inclusion in the Yellow Book.

**The genius of** Holland's tunnel was in its three tiers: the bottom is for supplying fresh air, the middle tier (*man standing*) is for two lanes of traffic, and the top is for removing exhaust.

**Meeting the** Interstate System design standards of "flat and straight" meant going through the Rocky Mountains, not over them. Seventy-five percent of the rock dug out for the Eisenhower Tunnel was granite. In the 30 years since it opened, over 200,000,000 vehicles have passed through the mountain.

# THE EISENHOWER MEMORIAL TUNNEL

Ike would be proud to have this tunnel carry his name. The Eisenhower Memorial Tunnel in the Rocky Mountains of Colorado is the highest vehicle tunnel in the world at 11,155 feet above sea level. It's also the highest point on the entire 42,795 miles of Interstate System.

Before there was an Eisenhower Tunnel, there was the Loveland Pass, named after W. A. Loveland, who tried but eventually gave up on building a railroad tunnel through the same mountainous terrain in 1867. This treacherous stretch of road was the only way to get over the Rocky Mountains when driving from Denver. Also known as U.S. Route 6, the pass was notorious for its avalanche chutes and was closed to all traffic during the winter months. Three years before the Eisenhower Tunnel opened, a stretch of U.S. Route 6 was identified as the deadliest stretch of highway in the United States. The locals said even its flat stretches were dangerous.

Engineers decided in the early 1960s that the design standards of the Interstate System could not be met by going over the mountains. The highway would be too steep and its curves too sharp. The only way to meet the network's high standards was to go through a mile and a half of granite mountain.

Nine men would die while blasting their way through that granite mountain.

## THE ILLITERATE MOUNTAIN

In December of 1964, workers "holed through" the mountain while drilling a pioneer bore. Barely large enough to crawl through, the minitunnel would give tunnel engineers an idea about what they were in for when digging the actual highway tunnel. Or at least that was the theory. One of those engineers later grumbled, "We were going by the book, but the damned mountain couldn't read!" It took nearly twice as long as planned to complete the tunnel's first bore.

Working overtime to catch up with the schedule, 1140 tunnel workers worked three shifts a day, six days a week for three years. The first bore and its two lanes were dedicated to the late President and father of the Interstate System on March 8, 1973. The following six years, while the second bore was being completed, traffic flowed through the tunnel, one lane each direction.

Today the Eisenhower Tunnel is one of the safest stretches of roadway in the world and has been credited with opening up western Colorado. The Interstate System has brought commerce to a formerly remote area of the state, and tourism has boomed too. The tunnel is full of skiers in the winter and vacationers in the summer. In fact, the tunnel is the only one in the world to pass beneath a ski resort. It is also one of the few places where a driver can enter one end of a tunnel in the midst of a blizzard and pop out of the other side to blue skies and sunshine. It's worth a trip to the top of America just to drive through this legend.

**In 1970,** Janet Bonnema's job application was processed by mistake. Highway officials thought she was a he. Miners believed the tunnel would collapse if a woman walked into it. Despite the men's superstitions, Bonnema eventually was accepted by them, but only after every worker had first walked out in protest.

## THE FORT McHENRY TUNNEL

In the 1970s, another plan for building a bridge turned into the construction of a tunnel. An eight-lane bridge was designed to carry the Interstate System over Baltimore Harbor in Maryland, filling in the final gap on America's busiest superhighway, I-95. The need for the highway was never an issue, as the missing link was a notorious traffic bottleneck. The debate centered on whether the pathway should go over or under the water.

Bridges, remember, are far cheaper than tunnels to build. The Federal Highway Administration, on the hook to fund 90 percent of the costs of the

*197*

**Fort McHenry** is famous as the setting of Francis Scott Key's "Star-Spangled Banner." To protect this national shrine, it was decided that a tunnel, not a bridge, should complete the nearby missing link in I-95, the world's busiest highway.

harbor crossing, was naturally pushing for a bridge. The problem was that the bridge would pass near historic Fort McHenry, the only national park that is also a national monument, the very place where an amateur poet named Francis Scott Key stood on the deck of a ship and watched "the rockets' red glare" in 1814.

The fort earned its place in history during America's War of 1812 with the British. Key was so moved during the enemy's 24-hour bombardment of the fort, and the Americans' fierce determination not to surrender, that he penned the famous poem. It was titled "The Defense of Fort McHenry," and it soon became a popular song, renamed "The Star-Spangled Banner." In 1931 the song became our national anthem.

Residents and historians were outraged that such a sacred landmark might be overshadowed by the noise, pollution, and unsightliness of a modern superstructure. They demanded that a tunnel be built instead. Besieged highway officials found themselves in the midst of a reenactment of the Battle of Baltimore Harbor. Residents donned garb from the period of the war and protested to TV cameras, decrying the desecration of Fort McHenry and the park around it

The bridge plan soon collapsed. The protestors, wrapped in their Star-Spangled cause, easily won the second battle of Fort McHenry. A tunnel it would be.

# BUILDING A TUNNEL—SAVING A FORT

When work began in 1980, the Fort McHenry Tunnel was billed as the "largest single project in the history of the National Interstate and Defense Highway Program." Its 1.7-mile tunnel was to take five years and $825,000,000 to build.

Instead of boring a tunnel through the earth beneath the harbor, a massive excavator vacuumed out a trench nearly 200 feet wide and about a mile and a half long. In that trench, 32 sections of prefabricated tunnel, each weighing 31,882 tons, were placed in two rows. One row, 16 tunnel sections making up four I-95 northbound lanes, was placed just 10 feet from the second row of 16 tunnel sections, four I-95 southbound lanes. It was the widest underwater tunnel in the world when it was finished.

While building the tunnel, a can-do spirit prevailed over and over again. Historians worried that artifacts from the famous Battle of Baltimore Harbor

**Built and launched** like a ship, this section of tunnel is about to make a 56-mile trip to Baltimore Harbor, where 32 of these sections were sunk and used in the building of the Fort McHenry Tunnel. Each section was 350 feet long and 83 feet high.

**The U.S. military** asked that the Fort McHenry Tunnel's ceiling be two feet higher than planned. Baltimore is a major port for military deployment during mobilizations, and the extra height allows military convoys to rush from nearby forts to the port. Here, emergency systems are put through their final checks before the tunnel is opened in 1985.

*Inset:* The tunnel was the largest underwater tunnel in the world when it was built.

**Not a single blind** spot: Highway officials monitor every foot of the massive tunnel with cameras and sensors buried in the roadbed for security and safety. If an emergency develops, police, fire, and rescue units are notified automatically.

*Visitors to the Fort McHenry National Park may never know that just 100 feet away and about 100 feet under the water, nearly 200,000 cars and trucks are roaring along I-95 everyday.*

might be lost forever during tunneling operations. Project officials called on the navy, which obliged by sending a minesweeper through the area to detect any underwater metal objects that might be of historical interest. They turned up only grocery carts and chunks of metal cable, and work on the tunnel carried on.

The military said the tunnel ceiling height of 14 feet wasn't high enough. They needed the tunnel to run at an extreme height of 16 feet so army transport vehicles could easily pass through during mobilizations. The project team met national security concerns and added the extra height.

It was, in the words of Ken Merrill, the tunnel's construction project manager, "a project that just made you feel good. The public was excited about it, and everyone who worked on it got along. It just worked out well, you know, on time—on budget!" Actually, it came in under budget when it opened on November 23, 1985, at a cost of $750,000,000.

Today, drivers may not realize that each one of the tunnel's 8,000,000 ceramic tiles was placed by hand or that 24 fans in two buildings are supplying them with fresh air and discharging polluted air. But far more important, visitors to the Fort McHenry National Park may never know that just 100 feet away and about 100 feet under the water, nearly 200,000 cars and trucks are roaring along I-95 everyday.

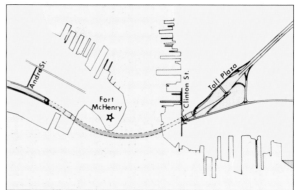

**Slaying** the Green Monster: The Big Dig's 10-lane tunnel replaces the elevated highway, known as the Green Monster, that has divided Boston's downtown from its water and its neighborhoods for nearly half a century.

# THE BIG DIG

Ike's Grand Interstate Plan ends with the Big Dig, in Boston. Also known as the Central Artery/Tunnel Project, the Big Dig is home to the most complex system of tunnels ever built. In a nutshell, it is replacing and expanding part of Boston's outmoded Interstate with state-of-the-art highway miles. Forty-two lane-miles of that system are underground, making up a series of tunneling marvels.

Boston's Central Artery, a section of I-93 known as the Green Monster, is one of the most traveled highways in the nation, hosting 200,000 vehicle trips a day. During rebuilding, the road could not simply be shut down with a sign saying come back in 10 years. So it was decided early on that the entire elevated highway would be held up on temporary supports while an enormous super-highway tunnel, 10 lanes wide, was beneath it. It's the first time in the world that a structure as enormous as the mile-long elevated I-93 has been fully under-pinned while an even larger piece of infrastructure is constructed beneath it.

The first step was to clear a path through the nearly 400-year-old city for the subterranean superhighway. Removing centuries of debris—colonial arti-facts, sunken ships, old wharves, glacial boulders, and ancient and unstable landfill—was the first step.

The last major obstacle was a 300-year-old maze of underground utilities. Decrepit phone lines were dug up and upgraded with fiber optics, then reburied. Abandoned pneumatic pipes for sending messages across town in the 1920s were exhumed and discarded. Water lines from the Civil War era—

*Left:*
**Temporary supports** are jacked into position under the elevated highway. Resting on the slurry walls that reach 140 feet into the earth, the new supports hold up the highway's daily load of 200,000 vehicles.

*Above:* **In order to keep traffic flowing** without interruption, one column at a time is "underpinned." Before the Big Dig, this type of operation was conducted for emergency repairs only. The Big Dig took the procedure to extremes by underpinning a mile of highway.

*Left:* **The old column** is cut with a welder's torch, and the weight is transferred from the old to the new supports.

*Right:*
**Custom designed** in Germany, this stout but powerful low-clearance crane dug the deep sections of slurry walls under the elevated highway.

*Below:*
**Smaller** and more nimble machines were used to maneuver around braces holding the slurry walls in place 12 stories below Boston's streets.

simple wooden tree trunks with five-inch holes burrowed through them to carry water—were still functioning and had to be rebuilt and relocated. In the end, over 31 utility companies spent six years clearing a path for the new tunnel.

## UNDERGROUND IN DOWNTOWN

The Big Dig's I-93 tunnel is unique for what it is not doing. It is not going under a river, under a harbor, or through a mountain. It is simply going underground in downtown to avoid going through town above ground.

Deciding to bury the highway was easy; doing it was another story. Building the tunnel required the world's largest use of slurry-wall technology—underground walls built to hold back the waters of Boston Harbor, allowing the massive tunnel to be built in a relatively dry environment.

By 1993, work on the slurry walls was under way. It would take a decade to finish the job.

The five miles of wall were built in nine-foot increments called slurry panels. Custom-designed machines from Germany, France, and the United States dug in the panels to a depth of 140 feet, deeper than the walls of the Hoover Dam penetrate Nevada's Black Canyon.

The slurry walls perform three major roles: They hold back the earth so the highway tunnel can be excavated between them. They hold up the existing Interstate, as temporary supports under I-93 are jacked up under the highway with the slurry wall as a foundation. Finally, in a first-time-ever application, the slurry walls became part of the final highway tunnel. Typically, slurry walls are used only as temporary support structures and are not incorporated into the final design. But in the Big Dig, where tight space limited construction, the walls remain, with panels of ceramic tiles placed directly against them.

## THE WORLD'S BIGGEST

In a nation where "big is better," the Big Dig claims a lot of bragging rights. Not surprisingly, the largest and most complex tunnel system in the world has the largest and most

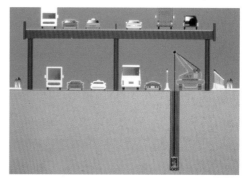

**Special low-clearance cranes** dig 140-foot-deep slurry walls under the elevated highway.

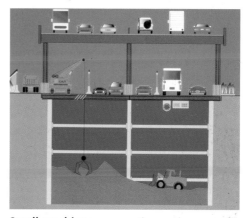

**Small machines** remove the earth between the walls, which now hold up the elevated highway.

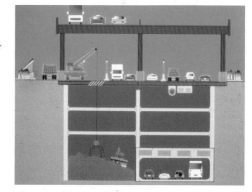

**The northbound tunnel is opened** a year before the southbound. Northbound traffic is taken off the elevated highway.

**Bound together,** two 140-foot-long steel beams called soldier piles are lowered 140 feet into the earth. After they're positioned below ground, concrete is poured around them, making a concrete-and-steel slurry wall panel nine feet in length. Five miles of slurry walls were built by completing one nine-foot panel at a time.

The Big Dig's first tunnel opened on December 15, 1995 and was named the Ted Williams Tunnel after Boston's legendary baseball slugger. At a cost of $15,782 an inch, it is one of the most expensive highways anywhere in the world.

complex ventilation system in the world. In just three minutes, the system can create hurricane force winds in the tunnel while replacing the air in all of its 42 lane miles.

Acting as the tunnel system's lungs, eight massive buildings spread around the city pull fresh air into the tunnels with enormous supply fans and pull it back out of the tunnel with equally large exhaust fans. One of the ventilation buildings has the deepest basement in the city of Boston, in order to reach the tunnel system.

The Big Dig's ventilation system finds its roots in Clifford Holland's tunnel. Throughout the project's tunnels are the three tiers for fresh air, traffic, and exhaust found in his Hudson River crossing. Little has changed in that regard. However, the mechanics of the Big Dig's ventilation system have been perfected with the latest scientific breakthroughs, including information from the West Virginia Memorial Tunnel program, described in the next section.

Today, the Big Dig's I-93 northbound tunnel is open, carrying drivers on a

*Inset:*
**It was tedious work** relocating 400 years of debris and utilities for the underground superhighway.

*Above:*
**Soldier piles,** looking like soldiers at attention, can be seen in the wall to the left. Wall tiles are eventually placed over them.

*207*

**The Big Dig has the largest** tunnel-ventilation system in the world. Eight ventilation buildings, like Vent Building No.7 at left, are placed throughout the city, pulling fresh air into 42 lane-miles of highway tunnels while pulling exhausted air out. *Above:* Huge air-intake vents. *Below:* Ten blue fresh-air intake fans and 14 yellow exhaust fans are used in Vent Building No. 7.

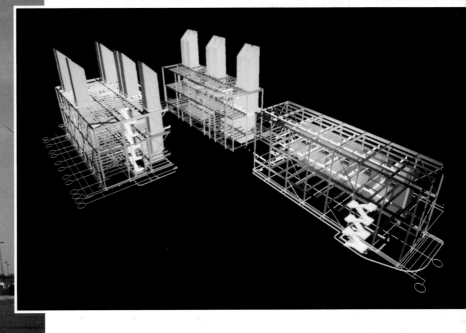

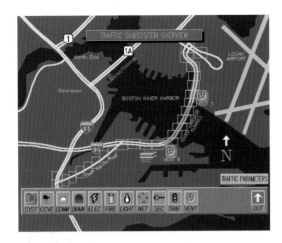

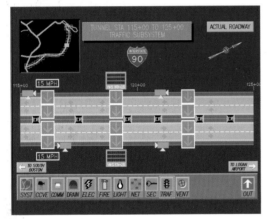

*Above right:*
**The most advanced** operations control center in the world monitors the tunnels of I-90 and I-93.

*Above:* **Computer** screens show stalled cars, accidents, fires, and even possible terrorist activity.

*Opposite:* **Subterranean** Vent Building No. 3 shows only its 270-foot-high exhaust stacks. Eventually a commercial building will be "wrapped" around them.

subterranean roller-coaster ride beneath the streets of Boston. Heading north into the city, drivers rocket down a steep decline when they first enter the new superhighway tunnel, which dodges a subway line that was built in 1916 at 80 feet underground. Still in the tunnel, the drivers climb over a second subway line at the tunnel's shallowest point, where the superstructure's roof girders are just three feet from the surface.

Taking one last dive into the underworld, reaching a depth of 85 feet, drivers on the I-93 northbound tunnel enter an interchange unique in the Interstate System—a completely underground intersection of ramps and highways connecting three separate tunnels. Within seconds, the underground voyage ends as the northbound lanes abruptly deliver car and driver onto the widest cable-stayed bridge in the world.

When the tunnel first opened in 2003, state police were forced to place a detail at its exit as at least six accidents caused by gawkers were recorded. The troopers' job, with lights flashing and hands waving, was to keep the motorists moving along. Many of the drivers were practically stopping in the tunnel and on the bridge, snapping pictures of the new Interstate superstructures.

*In 1990, crews unlocked the decommissioned tunnel, its road surface still showing its double yellow line, and started making $38,000,000 of improvements that were soon to go up in smoke.*

## THE MEMORIAL TUNNEL

The world never knew exactly what happened during a raging inferno inside a highway tunnel until the West Virginia Memorial Fire Ventilation Test Program. Before these unusual pyrotechnical experiments, any layperson could have told you that a tunnel fire was deadly and dangerous. But the scientific community needed to know more. What were a fire's characteristics when it was burning out of control in a large tunnel? How could it be controlled, or could it?

There were theories, hypotheses, and computer-generated models; but no government in the world had ever set a tunnel on fire and studied the results. In 1970 the American Society of Heating, Refrigeration and Air-conditioning Engineers began to lobby the United States government to do just that. Money and an available tunnel to burn had kept the program from getting underway; but thanks to the Highway Trust Fund and Interstate System, the association got its wish.

The money dilemma was put to rest when the Federal Highway

Administration said it wanted to explore the use of a developing trend in tunnel ventilation: jet fans. The feds wanted to subject the fans to the ultimate challenge—real fires in a real tunnel—before installing them on the Interstate System.

The problem of finding a tunnel was resolved when the Memorial Tunnel in the mountains of West Virginia was abandoned in 1988, after improvements to I-77 negated the need for the structure. In 1990 crews unlocked the decommissioned tunnel, its road surface still showing its double yellow line, and started making $38,000,000 of improvements that were soon to go up in smoke.

Becoming the largest employer in the mountainous region for the next five years, the project began ripping the half-mile-long tunnel apart. Crews set to, removing the old 1953-era ventilation system and replacing it with the most advanced systems available. They gouged a trench in the tunnel's old roadway, laying pipes to carry fuel to four large steel pans. The pans, eventually filled with home heating oil, became the epicenters of 99 test fires. Lying alongside the pipes was an electronic "raceway" that held communication lines to relay scientific data about the fire to computers outside the tunnel's entrance.

## FIRE IN THE HOLE!

In May of 1993, the first fires were set, and for nearly two years they kept on burning. First, small 10-megawatt fires were ignited until the largest of the fires—a 100-megawatt conflagration—burned, reaching temperatures well over 2400 degrees Fahrenheit, a level equal to the temperature the space shuttle experiences upon reentry into the earth's atmosphere. Closed-circuit TV monitors in fireproof encasements were positioned in the tunnel, similar to the ones NASA uses to record a rocket's liftoff. This gave the world a permanent visual recording of the hellhole's fury, an important feature for those who would make future decisions about their own tunnel systems and emergency response procedures. After all, seeing is believing.

On a clear day in July of 1994, "the big one" was torched, setting off the program's most intense blaze, a simulation of a tunnel disaster equal to that of

**Putting research** to use: Before the Big Dig's I-93 tunnel opened, the Boston Fire Department demanded a test of the ventilation system, which had been designed based on results from the West Virginia program.

A worker stands in a vent stack used to divert smoke from nearby I-77. **Fire protection,** *above right*, was installed in the tunnel just in case.

*Right:* **Pans for holding** the fuel to be ignited are in the center of the tunnel. Notice the faucets used for filling them with fuel from a remote location.

*Above:*
**A fire test pan** is set ablaze outside the tunnel.

*Left:*
**In the tunnel,** it was filed with increasing amounts of fuel in order to simulate bigger and bigger fires.

an exploding fuel-tanker truck. Tall meteorological towers, installed to keep an eye on exterior weather conditions and guard against a blaze spreading to the surrounding woodlands, had relayed good news. Wind conditions were ideal and it was okay to proceed.

An operator pushed two buttons. The first withdrew hundreds of gallons of oil from a large tank outside the tunnel and into the four large oil pans positioned 700 feet into the tunnel. The second button sent an electronic spark into the fuel, setting it ablaze.

Almost immediately, closed-circuit cameras picked up a torrent of heavy smoke barreling down the tunnel at an alarming rate, nearly 1000 feet a minute. An unlucky person caught inside would have had to sprint a four-minute mile to outrun the pursuing cloud. Before going black because of the dense smoke, the special cameras documented how the smoke maneuvered around trapped vehicles. Large steel cutouts simulating the automobiles and trucks stopped behind

**Jet fans** were permitted for use in the Interstate System's tunnels only after West Virginia fire tests proved their reliability under extreme conditions.

The highest elevation on the Interstate System is in a tunnel, the Eisenhower Memorial Tunnel in Colorado. The Eisenhower Tunnel also claims the highest elevation of any vehicle tunnel in the world, and it is the longest bored tunnel on the Interstate System, at 7789 feet.

the fictitious overturned fuel tanker were placed throughout the tunnel. It was a stunning display of how debilitating smoke reaches a trapped motorist long before the heat and flames do.

Estimating the time it would take an operator in the real world to respond to fire alarms, the tunnel's main fans were kicked on two minutes after the fire was set. Pulling massive quantities of the toxic cloud out of the tunnel, the fans in the fan room struggled in an environment hot enough to bake chocolate-chip cookies. The exhaust units, nonetheless, continued to operate as the temperature reached 400 degrees. Special stacks shot the smoke straight into the air to avoid blinding drivers on the nearby I-77.

Inside, as temperatures climbed to the 2400-degree mark, old bricks in the road exploded and ceiling tiles ricocheted off the highway-tunnel walls. A large tank of cold water outside the tunnel sent its contents, by way of pumps and pipes, to the special cases holding scientific equipment gathering data, showering the gear in an attempt to prevent its overheating and malfunctioning. A back-up power grid and fire-extinguishing system were ready to kick in if needed. The elaborate system succeeded, allowing computers to continue plugging away, gathering wind speeds, temperatures, gas emissions, and other critical information from inside the hellhole.

The turbines of six experimental jet fans attached to the ceiling of the tunnel, similar to the engines affixed to the wings of commercial airliners, began screaming in high-pitched tones, sucking the smoke and heat from the fire and directing it out of the tunnel vertically. All six jet fans were placed directly over the burning oil pans, subjecting them to the most intense heat possible.

The fans performed brilliantly, pulling the most ferocious elements of the fire—its intense heat and smoke—away from imaginary firefighters and rescue crews. As the heat from the 100-megawatt conflagration raged, the tunnel engineers watched through cameras, wondering how much more the fans could take. Inside the tunnel, high-strength stainless steel was melting like wax on a burning candle; but the jet fans kept performing. Only one failed.

**Today,** the Memorial Tunnel is being used to train Special Forces in antiterrorism warfare. It was Major General Allen Tackett's idea to lead his elite National Guard forces in simulated terrorist attacks in highway tunnels, parking garages, and subways.

The tests cleared the way for the use of the previously unproven fans, saving future highway projects, including the Big Dig, untold millions. On the Big Dig, the test program allowed for a partial reduction in what was already the largest ventilation system in the world. Saving $20,000,000, the tests also allowed the elimination of unnecessary and expensive fire retardants from the ceilings of the Big Dig's tunnel.

More important, the Memorial Tunnel fire tests promise to save lives in tunnel calamities around the world, as fire and rescue crews now understand much more about the mysterious and deadly ways of smoke and fire.

Today the tunnel continues its important role of testing and training, this time in the battle against terrorism. Outfitted with burned-out automobiles, subway cars, and concrete rubble, it has become a commando-drilling facility for the country's elite Special Forces. Once trained inside the Memorial Tunnel, these brave fighting units can be quickly deployed anywhere, whenever they are called.

CHAPTER ELEVEN

# The Road Ahead

"**THE INTERSTATES AFFECT** every American, every single day. It doesn't matter if you ride the subway or drive a car or live in the city or the suburbs. It doesn't matter if you eat red meat or tofu. This superhighway impacts your life every day!" The words are from Richard Weingroff, the Federal Highway Administration's historian. "Forget the Internet," Weingroff says. "This is the real superhighway; it makes real connections to everything you need, every day."

The "mighty network of highways" planned by Congress and President Eisenhower was supposed to be finished on October 1, 1972. That was the day the legislation decreed that the much anticipated road trip across the country, the one without a single traffic light, was to be a reality. In fact, the System was only three quarters complete on the scheduled date. But the unfinished Interstate System was already helping to make the American Dream a reality for millions.

By the mid-1970s, for the first time in American history, more people were living in the suburbs than in the cities or the countryside. To say that the new highways were responsible for delivering—lock, stock, and barrel—a house, a porch, and a patch of grass to everyone in the country would be going too far. But the Interstate System went a long way to meet the growing demand for a suburban lifestyle.

219

*By the 1980s, the Interstate System had turned an office, retail, and warehouse boom into a new suburban lifestyle.*

At the same time, the Interstate System was expanding Americans' parameters. Families were taking vacations farther and farther from their homes along its ever lengthening routes. The Interstate System was shrinking travel time between cities and suburbs while it was opening parts of the country that were previously unreachable. Businesses continued to take up residence in new office parks along Interstate highways and beltways, leaving behind the expensive and often run-down cities.

By the 1980s, the Interstate System had turned an office, retail, and warehouse boom into a new suburban lifestyle. More workplaces were now located outside of the cities than in them. In the decades since it began, the Interstate System has emerged as the most powerful tool of our domestic economy. It is a rolling horizontal marketplace touching every corner of the nation.

## BUSINESS BOOMED

One place that boom is most evident is in the southern part of the nation. Before the Interstate System cemented the New South together, a jambalaya of state and local routes haphazardly shuffled drivers between southern cities. Work was unskilled and hard to find; wages were low. The South was in a long battle with poverty, and poverty was winning. Then I-85 began rolling into people's lives, running from Virginia to Alabama and bringing a boom of trade that would change lives forever.

Even before the southern portion of the Interstate System was substantially complete, a rush of businesses looking for cheaper labor and land had begun sweeping through the South. Many of the businesses were moving down from the Northeast, leaving behind expensive union wages and high rents. I-85 made cheap land and lower wages even more accessible. Compared to the North, the Old South was a bargain. And for the workers, the new companies represented a bonanza. The incoming industries were able to offer southern laborers twice or even three times the pay they normally received.

Before the Interstate, the South's poverty meant that its best and brightest often left for the big cities to the North. The nation's rural population stopped declining when companies began to set up their operations at the on- and off-ramps along remote stretches of the Interstate System.

The boom also created an unforeseen problem: where to get help. Workers were nearly impossible to find in the early rush, and businesses went to extreme measures to staff their offices and assembly lines. "We're recruiting at the prison gates," said the president of one company. The desperate executive was looking for men becoming eligible for parole, but even that supply of workers soon dried up. Nothing daunted, the company tried to get Cuban

The longest route on the Interstate System is I-90, which cost $7,500,000,000 and runs 3021 miles from coast to coast. A few days after it opened, a mock funeral was held for the last traffic light to be removed from its path. The ceremony was complete with coffin, prayer, and over 1000 mourners. There was not a wet eye in the crowd.

refugees from Miami, with no luck, then began going to the army looking for 4-Fs. Some small businesses closed for lack of help, and cattle ranches suffered as cowhands walked off their jobs to the newly opened textile mills along I-85.

The Interstate System also had another powerful effect on the South. Not only was it responsible for an economic boom, but it was helping to usher in a new era of civil rights.

Integration in the workplace suddenly became the necessary order of the day. Every employer surveyed for a major newspaper story during the summer of 1966 said it was gladly offering blacks more opportunities for advancement than ever before, better housing, and increased pay. Some companies were even offering hospital benefits, a luxury never before seen by some workers.

Though employment managers were having a hard time because of the labor shortage, one noted, "One thing that's happened is the color line has been broken down, perhaps forever, or at least as long as the boom lasts. The pressure has brought integration in the mills and brought it in fast."

Walter Harper of South Carolina's development board made it official when he said, "Economics has brought changes in racial attitudes that the law couldn't. Economics has solved a lot of age-old social problems"—economics made possible in large part by the Interstate System.

## TOBACCO ROAD TO AUTOBAHN

And the best was yet to be. In the 1990s, after a long search, BMW, the German auto manufacturer, chose Spartanburg, South Carolina, as the site for a new $400,000,000 assembly plant capable of producing the brand's popular 3 Series high-performance automobiles. The site selection had been a brutal process lasting three years and involving 10 nations and 250 potential locations.

**Interstate System** access was so important to BMW that South Carolina's governor went door-to-door asking 140 homeowners to accept relocation offers.

Spartanburg had been BMW's favorite location from early in the process. There was only one problem. The perfect site, 1000 Spartanburg acres, had a development with 140 middle-class homes smack in the middle of it. But state and local officials were so determined to bring the automaker to South Carolina that they regarded this as a mere inconvenience. And after a $36,000,000 game of "Let's Make a Deal," the homes were bought and bulldozed, and the site secured for the town.

# Mom, Are We There Yet?

**S**AY THE WORDS "family vacation" or "road trip," and a handful of images comes to mind: miles of Interstate; squirming kids in the back seat of the minivan, those golden arches (at last!); the orange-roofed motel and the courtyard pool. Now common ingredients of our daily lives, the fast-food restaurants, the motels, and the family-friendly service are the work of a few entrepreneurs who transformed the way Americans eat, shop, and sleep on the road.

It all started with an ice cream cone.

In 1925, Howard Dearing Johnson borrowed $2000 to buy a corner drugstore in Wollaston, Massachusetts. Though he also sold candy, newspapers, and medicine, Johnson soon realized that the real action was at his soda fountain. Recognizing a trend when he saw one, Johnson set out to create the world's best ice cream cone. Fifty years before Ben or Jerry, he tinkered with his mother's recipe to make an all-natural, superrich, ultra-premium ice cream, selling for five cents in a cone. And the world—or at least Wollaston—beat a path to his door. By 1928 Johnson had created 28 flavors, and ice cream sales topped $240,000. The Howard Johnson story had begun.

Howard Johnson was one of the United States' great entrepreneurs, with that storybook combination of penniless beginning, extraordinary hard work, and impeccable timing. By 1940, he had 170 restaurants, most of them franchises, located along East Coast highways. He left nothing to chance. He designed the space and set the standards for the food and ice cream served in his restaurants. His "Howard Johnson Bible" dictated regimens for cleanliness, service, recipes, and menus. Even waitresses' appearance and rules of courteousness—such as how to assist customers with their coats—were strictly enforced. America's new "Host of the Highway" decreed that customers must always find a friendly welcome and reliable experience when they visited a Howard Johnson's.

Ever the trend-spotter, Johnson had noticed Americans' growing love for their cars and had attentively followed the talk about the proposed Interstate. By the time President Eisenhower signed the highway act into law, Johnson had 400 restaurants in 32 states. His company also expanded into motel services, providing dependably clean rooms to travelers at his typically good prices. The rooms were even equipped with that new invention: television.

The car-based, travel-easy Interstate world was exactly what Johnson had been waiting for. By the time he handed the company to his son in 1959, people all over the country knew that the big orange roof meant friendly service, good food, and lodgings at a fair price. It had become one of the most familiar icons in America.

## HOLIDAY INN

The Johnsons were not the only ones watching events along the Interstate. On a family vacation, Tennessee entrepreneur Kemmons Wilson found the motels uncomfortable, overpriced, and unreliable in their quality. Not one to miss an opportunity, Wilson with his partner, Wallace Johnson, opened the first Holiday Inn on the outskirts of Memphis on August 1, 1952. Named after the Bing Crosby movie, the motel offered 120 guest rooms, each with its own bathroom, air-conditioning, and telephone. The complex also featured a swimming pool. Children stayed free in their parents' room.

Wilson promised his dubious wife that he would open 400 motels by the decade's end, and he did. In fact, within 20 years he had franchised over 1400 hotels. He had three times as many beds as his closest competitors, leaving even Howard Johnson's in the roadside dust. Like Johnson, Wilson insisted on consistent quality, affordable prices, and highway convenience. Known for choosing locations by airplane at proposed Interstate exchanges, Wilson made sure the 43-foot HOLIDAY INN signs were easily visible from the road, and the rooms themselves were easy to reach from the family car.

# McDONALD'S

The McDonald's story began with a road trip and a milk-shake machine.

Ray Kroc, a 52-year-old Multi-mixer milk-shake-maker sales-man, drove west in 1954 to check out two brothers who were reportedly using eight multimixers at once. This busy an enterprise the salesman had to see with his own eyes—a now legendary decision.

The brothers—Dick and Mac McDonald—ran a hamburger stand in San Bernardino, California. They had grown tired of having to replace short-order cooks, waitresses, and broken glassware, so in 1948 they retooled, reducing their menu and eliminating anything that required a knife or fork. The brothers then applied Henry Ford's assembly line principles to their kitchen, separating food preparation into simple, repeated tasks. In search of a broader family clientele, they then hired young men only, fearing that female workers would attract teenage boys who in turn would scare away everyone else. The brothers' new hamburger stand became a great success, with a growing national reputation.

When Ray Kroc saw how profitably the brothers ran their restaurant, he immediately suggested that they go into business together. By the end of John F. Kennedy's first year in office, Kroc bought out the brothers and set about building an empire, locating his restaurants near Interstate interchanges and emphasizing tasty hamburgers with his special sauce, low prices, and cleanliness. "If you've got time to lean, you've got time to clean," he told his employees.

Kroc's McDonald's restaurants weren't the first drive-in franchise hamburger stands. That honor belongs to White Tower. But McDonald's easily and quickly became the largest. Today, there are over 30,000 worldwide, and the golden arches are one of the most recognizable structures in America, a beckoning beacon to travelers.

In the car-happy, Interstate-loving late 1950s and '60s, Kroc inspired a great range of imitators. Those whose companies are still around include William Rosenberg (Dunkin' Donuts), Glen W. Bell, Jr. (Taco Bell), Keith Cramer and Matthew Burns (Burger King), Dave Thomas (Wendy's), and Harland Sanders (Kentucky Fried Chicken). There were even more enterprises that didn't make it, such as Burger Chefs, Burger Queens, Burgerville USA, Yummy Burgers, Twitty Burgers, and Burger Boy Food-O-Ramas.

None of these businesses would have worked without the Interstate. Not only did the System provide customers, it also made the franchises practical. The highways meant, for example, that beef providers could be located next to feedlots and away from urban areas, cutting down expensive travel time from field to market. In a nationwide web, trucks shipped ingredients from field to processing to table wherever those tables were located. The McDonald's corporation is the United States' largest purchaser of beef, pork, and potatoes, and the second largest purchaser of chicken. Ease of shipping is critical.

Johnson, Wilson, and Kroc were the leaders of a drive-in nation. Virtually everything has been tried along the Interstate: drive in theaters, banks, shopping, college registration, even church. In fact, a church in California that offers parishioners the chance to "worship as you are . . . in the family car" has become one of the largest congregations anywhere.

The highest officials in the state and nearly every local politician had worked to entice BMW to the area. In addition, Spartanburg offered the BMW executives a wonderful climate, inexpensive labor—and most important—ideal access to the Interstate System. In return, the manufacturer brought 2000 wage-earning jobs, jobs important not only to the town but to the elected officials.

Spartanburg had been doing well even before BMW came on the scene. It was an early beneficiary of the Interstate boom, and it had banished its Tobacco Roads and shanties long before. But with the coming of BMW, Spartanburg is prospering as never before. So many German firms and others have followed BMW's lead in setting up operations along the superhighway that locals have nicknamed that stretch of I-85 the Autobahn.

The only problem is that I-85 brings too much of a good thing. Along some sections of I-85, traffic jams are choking out the opportunities. In Atlanta, the best areas for development are no longer found just along I-85. Now I-75 and I-20, two superhighways that also pass through Atlanta's center, are seeing phenomenal growth along their corridors. Wherever there is growth in the country, it is almost without question near an Interstate.

## SPRAWLING SUBURBS

It's an inevitable chain reaction. An Interstate highway, like I-85, is built and the traffic seeks out the new road. Leaving behind previously congested cities or shipping routes, drivers can increase their travel speeds and reduce commuting times. People discover this desirable locale and flock to it, expanding business-es and building new homes. The land in the area increases in value, but slowly congestion creeps in and the good is gone. The process keeps repeating itself: Another site is found, the road is widened . . .

The phenomenon is called urban sprawl. According to Merriam-Webster's dictionary, the term originated in 1958, before much construction of the Interstate System had begun. In fact, misdirected and random development has been going on for as long as there have been cities. The automobile and superhighway have only accelerated an old tradition of seeking out more desir-able areas to live.

Some people see the Interstate System as the root cause of urban sprawl, but blaming it for urban sprawl is like blaming the sun for skin cancer. In both cases, controlling exposure to the former should prevent the latter; but it's not that simple. Just as many sunbathers can't get enough sun, most Americans can't get enough of their Interstate System.

The good news is that we have a lot of land in this country. In the mid-1990s, one survey showed that if every household then in an apartment

*Opposite:*
**By the mid 1970s,** for the first time, more Americans lived in the suburbs than in cities; and by the 1980s, more offices were located outside cities than in them. Malfunction Junction, the Mixing Bowl, and Spaghetti Interchange are a few of the nicknames given to interchanges that lie outside cities. Shown here are the high-speed flyovers of I-85 and I-285 on Atlanta's beltway.

225

I-95 through New Jersey is an optical illusion. The town of Hopewell, New Jersey stopped I-95's construction in the 1970s. Today, the New Jersey Turnpike carries the I-95 Interstate System shields even though about 20 miles of its planned route between Trenton and New York City do not officially exist.

*The Interstate System is key to moving people and freight through and around every major American city.*

building or freestanding home was placed on one acre of land, only five percent of the lower 48 states would be occupied. The bad news is that development along the Interstate System is spreading like a fast-growing vine up a tree.

In the city of Phoenix, Arizona, along I-10, I-17, and other expressways, the city is growing at the rate of over an acre an hour! When the Interstate System was started in 1956, Phoenix's suburb of Scottsdale was barely a solitary square mile in which 2000 people lived. Forty years later, it was three times the size of San Francisco, with over 165,000 people.

Our cities keep getting bigger and bigger, mainly growing out along their Interstate Highways. In 1956, about 13 percent of the planned Interstate System was considered urban highway. Today that number has more than doubled. Twenty-nine percent of the Interstate System is now considered to be in a metropolitan area. The Interstates haven't moved; it's the cities that are sprawling.

Americans' right of passage—to move on, hit the road, pull up stakes, and make a claim on a greener pasture—comes at a big cost. Farmland is disappearing to developments at a rate of 1,200,000 acres of land a year. Still, we push out farther and farther from the city centers looking for bigger and more affordable homes with that backyard and two-car garage.

Today, nearly every state says that its Interstate System plays a critical role in urban travel. Almost 70 percent of every state's highway budget goes toward building, repairing, or maintaining its urban Interstate System. Half the states have found that the Interstate System has become the key connection between their downtowns and their airports.

Perhaps Chief MacDonald would say, "I told you so," if he were here today. He and his staff at the Bureau of Public Roads had forecast in the 1930s that the Interstate System's main objective would be facilitating the movement of traffic in and around a city. Despite the initial romantic notion of transcontinental travel, some 70 years later we are finding out that they were exactly right. The Interstate System is key to moving people and freight through and around every major American city.

## TRUCKIN' ON

All of this expansion along the Interstates must of course be provided with the goods it needs. Businesses need their computers and water coolers . . . and the bottled water that goes into them. Suburban homes need their appliances, the clothes to fill those walk-in closets, and the astonishing variety of food and drink that the American public takes for granted.

To bring those goods home to America, a new industry exploded onto the Interstate—trucking. In a direct lineage from the old Conestoga wagons,

**Trucks and tractor-trailer rigs** populate the American landscape. The Interstate System has emerged as the most powerful tool of our domestic economy, a horizontal market touching every corner of the nation.

tractor-trailers are a ubiquitous sight along the nation's highways. They weigh 40 tons, cruise at 70 miles per hour, and their drivers hate to stop. Big rigs have become an American icon, pushing their way onto the center of the nation's economic stage. More warehouse than vehicle, these powerful tools of trade move nearly everything we touch in the course of a day. Thanks in part to these rubber-tired behemoths, the United States now has the most efficient economy in the world. And its superhighways have become the new rivers of commerce.

Over the last 10 years, truck traffic on the Interstate System has increased by 42 percent. Over one million more trucks are on the highway now than there were 10 years ago. Ninety-three percent of the nation's freight is hauled over the country's highway network, and the Interstate System is the central component to that network.

# An Interstate Success Story: UPS

**U**PS **IS THE LARGEST** package-delivery company in the world, hustling over 13,000,000 packages between 8,000,000 customers everyday of the year. In 1975 it was the first company to service every address in America. Today it is one of the largest single users of the Interstate System.

UPS is a master of the express-carrier universe, watching over everything in its sphere of influence, from how long it takes the men and women in brown to read a package label (1.29 seconds) to how a driver should carry UPS truck keys (on the pinky finger). To say they have studied and perfected the use of the Interstate System is a gross understatement.

"The Interstate System is the backbone of our entire domestic operation," explains Dan McMackin of UPS. "At one point or another every package we handle probably touches or rides along on it. Whether a parcel is being shipped on a package truck from one side of a city to another or across the country by rail or even on one of our aircraft, the package, almost without exception, is going to travel on the Interstate."

## INTERSTATE COMMERCE

UPS, along with the rest of America's most established companies, has grown and prospered in step with the nation's highway system. Its corporate history mirrors that of the road.

In 1907 a resourceful teenager named Jim Casey was ready to improve his lot in life. After running bail money to imprisoned miscreants and cocaine and opium to hopheads, Casey decided enough was enough and founded what is now UPS. With a loan of $100, the budding entrepreneur set up a phone bank on a borrowed lunch counter in the basement of a Seattle, Washington, building and started a foot-and-bicycle messenger service.

Before long, Casey was running a Model-T along with his bikes. Soon he and his partners were in charge of a growing fleet of trucks, all painted Pullman brown—the same color as the prestigious railroad cars of the day. Besides being a color of status, it was great for hiding road dirt. Today, it's UPS brown.

Always ready to adapt, Casey sensed that after World War II, people and their automobiles were moving out of the cities and heading to the suburbs. Casey decided it was time to expand. This time he would challenge the monopoly the United States Postal Service held on parcel post.

Before the Interstate System was up and running, it was illegal for any operation except the United States Postal Service to deliver packages across state lines. UPS felt that in many cases it could deliver mail and parcels better and faster than Uncle Sam. By the 1970s, after many bureaucratic battles and victories, UPS was able to start taking advantage of the new superhighway for its business. Today, its brown trucks run everywhere the Interstate System goes.

## BEEHIVES AND BIG BROWN BEES

There are no coincidences at UPS. For example, every driver of the 78,814 UPS brown package trucks is instructed to pull the seatbelt over his chest with his left hand while inserting the key into the ignition with his right hand. It saves time, as he'll be delivering over 200 packages throughout the day. Also for the sake of saving precious time, most of UPS's 1748 large sorting-and-distribution facilities are very near to the Interstate System.

Some UPS drivers call these distribution centers beehives, making them and their trucks very large brown bees. In the morning the hive is swarming with activity.

At 7:30 a.m. in Shrewsbury, Massachusetts, Dick Socha, a driver of one of the company's UPS-brown tractor rigs, is under the hood checking its vital fluids. While he slept the night before, this truck, called a feeder truck, was in motion hauling goods back and forth from Albany, New York. The term feeder refers to the loads of cargo the truck carries, which are fed to the beehives up and down the Interstate System. Once a feeder pulls into a hive, its packages are unloaded, distributed, and reloaded into smaller package trucks for local deliveries. Feeders run 24 hours a day for up to 12 years, putting on a million miles, most of them over the Interstate System.

Socha, a 31-year veteran of the UPS army, washes the windows of his Mack truck, tops off its two 62-gallon diesel tanks, checks the pressure in its tires. After his feeder passes his inspection, his delivery day begins.

Today, Socha is hauling tons of packages to the Meadowlands, outside of New York City. He knows that the trip is about 420 miles and could take him as long as 12 hours round-trip. He will use six different Interstates, I-90, I-84, I-91, I-95, I-87, and I-287

**Before leaving the Meadowlands** in New Jersey, UPS driver Dick Socha confirms that the "pups" numbers match his paperwork.

before the day is done; and he knows he will probably see three or four accidents along the way. He always does, but he has 21 years of safe driving, so he's not worried. Besides, even I-95, the most heavily traveled highway in the world, is still safer than most other roads anywhere else in the country.

Despite everything UPS does to save time on Socha's trip, once he hits the road his travel time along the overcrowded highway is a gamble. "When 95 wasn't so congested, drivers were told to stick to it," recalls Socha. "Now drivers are allowed to pick the quickest way they can find, depending on traffic conditions. An accident, a breakdown, or a lane closure can ruin your day, so I listen to AM radio to get traffic reports every ten minutes."

Arriving at the Meadowlands, Socha drops off his large single trailer, exchanging it for two "pups" that came up from Baltimore. These smaller trailers are connected like boxcars on a train and require more of a driver's skill and attention, as they tend to swerve if the rig is not driven exactly straight. After lunch, Socha is on his way back. Taking a gamble, he heads north over the George Washington Bridge and through the Bronx on I-95. Who knows what traffic conditions lie ahead?

Playing the role of the winning tortoise, Socha is chugging along in the slow lane, doing the limit—

even letting other drivers cut in ahead of his feeder. At the end of the day, he will have beaten the hares by arriving at his destination in one piece and with peace of mind.

Before taking his Interstate off-ramp home, he will pass four accidents, one car fire, two rear-enders, and a fender bender. Along with every truck on the Interstate at the time, he stops at two weigh stations where officials are busy apprehending overweighted rigs and drivers who may have cooked their logbooks to hide excessive hours of driving. As several rigs are forced to stay behind and be ticketed, Socha's feeder is waved through.

Pulling back into the hive in Shrewsbury, Massachusetts, Socha's parks his brown bee and the two pups. A throng of other brown bees is swirling around the yard, checking in with dispatch, unloading, and refueling.

The twelve-and-a-half-hour day has been a successful one as Socha and his equipment have returned ready to do it again tomorrow. In a couple of hours, while he is sitting down to dinner with his wife, the feeder he drove will be heading back out onto the Interstate System.

**The nation's** economic might is on display on the Interstate System, weighing 80,000 pounds, decked out in chrome, packing 500 horsepower, and carrying 300 gallons of fuel. There are 1,000,000 more trucks on the road today than 10 years ago.

At the moment, 66 percent of the Interstate System's surfaces are in good, though not great shape. To maintain this level, the country must spend at least $23 billion on upkeep. Making substantial improvements in the System will cost a staggering $125 billion.

## ROLLING STOCK

Trucks and the superhighways on which they travel make for efficient shipping, cut down on costs, and help to keep in check the retail prices of nearly every consumer good. The American supermarket, to name just one retail outlet, is a wonder of the modern world, thanks in great part to the trucks that stock it with food. Walk down any produce aisle and count the kinds of lettuce alone. There are pineapples in winter and strawberries in autumn, fresh vegetables all year round. Delicacies that in another age were literally available only to royalty spill off the counters of the neighborhood supermarket. And they are relatively cheap, affordable to almost all Americans, who regard daily large portions of fresh food as a birthright. And indeed that's what it's become.

In order to deliver these goods, truck drivers have had to become efficient managers as much as movers of cargo. Synergy is working its magic as the real and virtual highways come together. Drivers use laptop computers connected to cell phone modems to receive real-time weather conditions and the most current road maps, complete with construction zones to be avoided. Laser printers download and print fuel prices per state, shipping routes, and pick-up and drop-off instructions.

Global Positioning Systems, known as GPS, allow dispatchers to keep tabs on their fleets of trucks and their cargoes. This last is a double-edged sword, as company truck drivers aren't always happy that their every move is recorded.

The boss knows if they're parked or making good time down the Interstate. But it makes scheduling much more accurate and keeps customers apprised of the location of their goods. In extreme cases, a GPS can save the life of a stranded driver who has veered off the road or is experiencing a medical emergency.

Soon, technology will allow truckers to use handheld computers to call ahead to a truck stop, select a meal, and schedule a shower, a refueling, and a parking spot for an overnight stay.

But there is a problem ahead. Today, almost half of our urban highways are considered congested. At the same time, the American Trucking Association estimates that 65 percent of all freight is shipped on those same crowded Interstate highways. And that number will only increase in the years ahead. Slower deliveries and increased shipping costs are a certain outcome, driving up the cost of our manufactured goods.

And it's not only trucks that need more and better roads. The number of day-to-day commuters and the distances they commute are growing steadily and show no signs of slowing down, any more than their cars do. Even with the trend toward at-home offices, the average commuter spends one and a half hours in his car these days.

The trick the nation must pull off is accommodating these drivers without simultaneously crisscrossing the landscape with intrusive, polluting highways. The highways of the future will have to employ innovative engineering—and social—expertise if our country is to remain clean and beautiful and safe.

Fortunately, the nation's road builders feel the same way.

Rest stops along the Interstate System appear on the average of once every 35 miles, so using the restrooms when you see them, not just when you need them, is a good idea.

## GLENWOOD CANYON: HARBINGER OF THE FUTURE

One of the most amazing pieces of construction on the Interstate System is the stretch of I-70 that passes through one of the most beautiful places in this country: Glenwood Canyon. Glenwood is a 2000-foot-deep canyon carved billions of years ago by the rushing waters of what we now call the Colorado River. It is a ruggedly magnificent world of steep, jagged cliffs chiseled out of granite and quartz. It was known to the Indians, who avoided it because of the difficulty of the terrain. But early pioneering trappers hacked a rugged path along its rocky outcroppings, and settlers saw it as a shortcut to the west.

In 1902, the first car bounced and sputtered its way through the canyon along a newly built road. That same road was widened and improved in the 1930s, much of it with dynamite, and became an official two-lane highway marked U.S. Routes 6 and 24. And when the Interstate was laid out in the '50s, a piece of the new highway was slated to pass right through its ancient cliffs.

Unlike the situation with other Interstate routes, I-70's engineers were

**Full-size model** columns made of plywood helped highway planners visualize what the real columns would look like.

**Contractors** were fined up to $22,000 per destroyed tree.

immediately challenged by environmentalists and other groups. Coloradoans wanted an Interstate, but they were also keenly conscious of the beauty of their state and the dollars it attracted in the form of tourism. They weren't about to mar a canyon that was worth its weight in visitors' dollars.

Setting aside their suspicions and differences, community activists, environmentalists, and highway officials put on hiking boots and trekked the road's projected 12.5-mile path through the canyon, determining exactly where and how the highway should move through the area. Colorado's lawmakers were clear in their demands, saying that designers needed to find a "blend between the wonders of human engineering and the wonders of nature."

During construction, both lanes of old U.S. Route 6 and 24 were to remain up and running; but after work was complete, Route 6 and 24 would be completely removed from the canyon, leaving a biking and walking path in place of the old roadway.

Interstate 70 would be four lanes of highway with a world-class design blending the highway into the canyon walls. Copying techniques used in building highways in the Alps, construction would be done from cranes hanging over trees and cliffs, avoiding the disruption of rocks and fragile ecosystems. Six miles of the 12.5-mile route were to be built on 39 bridges, viaducts, and tunnels, with over 15 miles of massive retaining walls.

And somehow it worked. In an unprecedented display of synergy, highway engineers became environmentalists, and environmentalists became highway engineers. As Ralph Trapani, the project's director, said in one interview, "Old, grouchy highway engineers [wound up arguing] to save a tree. Then the environmentalists would talk safety. It was the most complete flip-flop of culture I've ever seen."

Environmentalists and a watchdog Citizens Advisory Committee kept tabs on their canyon. For each raspberry bush contractors destroyed, they were billed $35; for each flattened scrub oak, $45; and if they were unfortunate enough to kill a blue spruce or cottonwood, they had to hand over $22,000. Wetlands that were taken were replaced by new ones. Full-scale plywood models of the highway's massive columns were erected to see what the real concrete and steel versions would look like.

Both contractors and highway officials set out to leave the place better off than they found it. Canyon walls that were scarred during the construction of U.S. Route 6 and 24 were sculpted to look as they did before the dynamiting and disfigurement. To make sure the newly revealed surfaces blended in, a dye was applied to the rock that matched the patina of surrounding canyon walls. Twenty years after their disappearance, two herds of wild Rocky Mountain bighorn sheep were reintroduced to the canyon. Over 150,000 new pieces of

*Left:*
**Looking west** through Glenwood Canyon. Indians avoided traveling through it. Settlers first passed through by navigating the Colorado River in boats. Today, I-70 hugs the canyon's walls with a specially designed safety railing that allows motorists to take in the view.

*Below:*
**The highway** project included reintroducing bighorn sheep to the canyon, where they hadn't been seen in 20 years.

233

**The Hanging Lake** Tunnels: Two of the 104 tunnels on the Interstate, they funnel traffic from viaduct bridges near the narrowest section of Glenwood Canyon.

To accommodate the military's need for higher clearances on the Interstate System, the standard minimum height of overpasses was set at 16 feet in 1960. Some older bridges have been modified to meet the standard, but many still do not offer the required clearance.

indigenous vegetation and trees were planted.

One big problem was that the highway was on a collision course with the head of the state's most popular hiking trail near Hanging Lake. To make matters worse, the trail was located in the narrowest point of the canyon. There was only one option—put the highway through the mountain, leaving a car-free zone for hikers, bikers, boaters, and horseback riders.

It was done, and in style. The Hanging Lake tunnels are two of the most extraordinary in the United States. Because they cut through high-quality stone, mostly quartz and granite, the tunnel's two 3900-foot-long bores are the only major tunnels in the nation able to support themselves without the aid of steel or concrete reinforcements.

234

Since the location of the tunnel is so difficult to reach, making emergency response a grave concern, a fully contained control center was built. If it were left above ground for hikers to see, that would defeat all that was achieved, so the facility was built underground . . . all four stories of it.

From this 40-foot-deep bunker, highway personnel use a $3,650,000 electronic system of cameras and detectors, ensuring that everyone passes through the mountain safely. The system even has the ability to interrupt a vehicle's AM/FM radio to inform occupants of an emergency. Twenty electronic variable-message signs tell drivers about ice, fire, and lane closures.

The Glenwood Canyon stretch of I-70 was completed in 1992 and has been chosen one of America's ten most beautiful roads by the AAA. The lessons learned there are still studied by engineers and environmentalists. Today, whenever a scenic site is in the way of a road, and the community and the engineers who want to build that road sit down to talk, their discussions profit immeasurably from the meticulous work done at Glenwood Canyon.

## SAVED BY SUPERPAVE

Along with beauty and engineering, the people planning our highways of the future are looking at the very surfaces on which those millions of cars and trucks will be driving. Three quarters of the Interstate System is paved with asphalt, as are 95 percent of all paved roads in the nation. But not all asphalts are equal.

At its most basic, the asphalt used in building roads is a mixture of petroleum by-products and gravel or crushed rock. But the combination and size of the rocks, the amount and consistency of the tar, and the recipe of other materials all can differ. What is the best combination? And does that "best" combination vary from climate to climate?

In 1987, Congress saw a disturbing trend in the world of resurfacing. Almost no money was being spent on research and development by the highway community. Roads were deteriorating at an alarming rate, and the costs to repair them were skyrocketing. Something had to be done to make the black stuff and what's under it last longer.

With $50,000,000, Congress supplied the muscle to develop the Superpave methodology, a practice by which odd bedfellows—academicians, contractors, manufacturers, and government officials—agree on how to mix, test, and apply asphalt, ensuring that your blacktop is all that it can be.

Before Superpave, the approach to asphalt design meant that Idaho's highway department might be mixing its asphalt in the exact same way as Arkansas or Maine, with little or no regard for temperature, rainfall, or traffic

**Working in** furious spurts over weekends, Georgia highway crews rebuilt eight lanes of I-285. Crews placed up to 31,500 tons of asphalt over 56 hours, using 225 dump trucks and 14 giant milling machines. Contractors were fined $100,000 an hour if they interrupted the Monday-morning commute.

volume. Now all 50 states agree that they each need something special and have agreed on how to determine what that something special is. That's Superpave in brief.

Today, asphalt surfaces that might have worn out in as little as 10 years are lasting 15 years or longer. Fewer potholes, repair crews, and delays on the Interstate System mean that more than 2.5 billion dollars a year is being saved.

## A HOT MIX

One of the things that makes asphalt so well suited for the Interstate System is that compared to concrete, it goes down quickly and is ready for use quickly. It minimizes the highway user's downtime, a prerequisite to all Interstate System jobs.

That quality was important in reconstructing 32 lane-miles of Atlanta's beltway, I-285. Originally built as a four-lane concrete highway and then widened to eight asphalt lanes in the early 1980s (before Superpave processes were devised), it was in bad shape and needed to be completely rebuilt. The good news was that the old asphalt could be reclaimed and remixed to Superpave specifications. Asphalt, in fact, is the most recycled material in the country.

But however fast asphalt is, it would still require some downtime. The choice in front of highway officials was to be cursed by the driving public while doing the work over 540 days with partial lane closures or to be cursed just as hard but for only 51 days with a total closure.

Instead of dragging out the process, Georgia officials decided to fast-track the job with a total closure of the highway on weekends. A gutsy call made by brave souls.

"It would not have been feasible to do it with concrete," explains Mickey McGee, one of the state's engineers on the job. "Concrete is a slower process, while asphalt is a much faster and simpler process. . . . I'd guess that if we had used concrete on this job, it would have taken three years to complete."

Biting the bullet, closing traffic and working furiously for 22 weekends, crews began their task in July 2001. Starting at 9:00 p.m. on a Friday, they hit the job running. Giant milling machines, as many as 14, began grinding and pulling up the entire 11 inches of surface on the worn-out superhighway. Regurgitating the torn up asphalt onto the side of the road, the milling machines, work crews, and support vehicles worked nonstop until Monday morning. If the work crews were not off the Interstate by 5:00 a.m., the contractor was fined $100,000 for each hour of delay. Not surprisingly, all eight lanes were ready for rush hour every Monday.

Drivers who accidentally veer into the path of oncoming traffic need at least a 30-foot buffer to regain control of their vehicles. As a result, Interstate System medians that do not have protective barriers are 30 feet wide and even wider when higher speeds and steep hills add to the danger.

*237*

As many as 225 large dump trucks a night ran convoys between the work-site and the plants producing the hot mix—the slang for the asphalt mix. The shredded asphalt was dumped into a recycle chute, and the old highway material was heated up to nearly 400 degrees Fahrenheit as new oils, additives, and aggregate were added per Superpave specifications, making the old stuff like new.

Still steaming hot at 300 degrees, the recycled asphalt was trucked back to the highway. There, at the rate of nearly 10 tons a minute, it was unloaded, spread, and rolled. Definitely saving time and perhaps the lives of drivers and workers (both were safer staying out of each other's way), the weekend-closure method worked so well that it's become a model for reconstruction jobs elsewhere on the Interstate System.

## PAVING THE HIGHWAYS OF THE FUTURE

Hidden in stretches of Interstates, sometimes marked with small blue signs but invisible to the cars whizzing over them, are test sections—pieces of roadway being monitored to check on how well the surface is holding up. These sections are watched carefully, but the real scientific fun has been on the test track.

**Finished in 1999,** Nevada's test track used driverless trucks. Computers in the sleeping cab drove the vehicles. After a million miles, one accident destroyed two computer-driven trucks.

In 1998, at a control room at WesTrack, an hour outside of Reno, Nevada, an operator watched four bright red tractor-trailer rigs, each pulling triple yellow trailers and weighing 150,000 pounds, chase each other around a 1.8-mile oval test track at 40 miles per hour. The weird thing was, no one was driving the trucks.

The operator in the control room focused on four terminals and their computers. Each computer was in command of one of the trucks, adjusting its steering wheel, throttle, transmission, brakes, and every other key component by radio. Every second the computer diagnosed 160 vital signs a trucker would normally look for, like gas, oil, and engine-heat levels. The trucks adjusted the steering wheel 10 times more often than a human would, keeping the trucks within two inches of their predetermined paths.

The trucks averaged 770 miles a day—equal to a trip from Chicago to Philadelphia—for two years, with no one at the wheel, and each traveled more

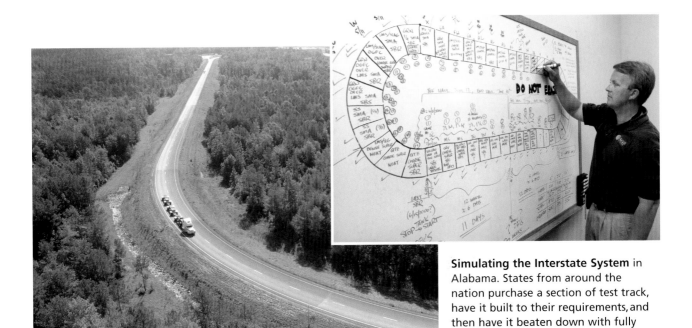

**Simulating the Interstate System** in Alabama. States from around the nation purchase a section of test track, have it built to their requirements, and then have it beaten down with fully loaded trucks. State officials follow the results on the Internet.

**The most advanced** gantry crane in the world places a 70-ton concrete viaduct section of a new I-93. Beneath it, an Acela train speeds to New York City.

I-495 around Boston Massachusetts, is the longest beltway on the Interstate System, running a 122-mile-long semicircle around the city.

than 500,000 miles, condensing 15 years of wear and tear on an average strip of Interstate System into 24 months.

The simulated road they drove around on is funded by the Federal Highway Administration and is designed to fail, at least once and in some cases twice. The 1.8-mile track is divided into 26 sections, each with its own type of asphalt mix. Some sections have Superpave mixes with larger stones as aggregate, and some have hybrids of concrete and asphalt. A few of the mixes were intentionally deprived of the recommended amounts of oil in their base to discover how fast poorly mixed materials would fail.

In 1999, when WesTrack's experiments came to an end, the federal and state governments maintaining the Interstate System had a much better idea of how to set standards for their road surfaces. They also had firsthand evidence of the damage contractors or manufacturers of hot mixes could cause if they violated production requirements. WesTrack helped determine what punitive fines would be levied against the delinquent and also helped to establish a reward mechanism for those who met and exceeded their obligations to the American public.

## THE BIG DIG

All of the research, testing, new engineering techniques, the community interactions and historic preservation methods developed during highway construction so far came together in one giant Interstate project, the Big Dig in Boston, which then added many new twists and discoveries of its own.

The Big Dig is the most complex civil project ever undertaken in the history of the United States. It's advancing civil engineering much as President John F. Kennedy's space program did aeronautical engineering in the 1960s. It's redefining how infrastructure, in particular the Interstate System, is being built and rebuilt. It's a model for the future, where the people most affected by the construction of a project are the ones with the greatest input into what gets built and how the building is carried out. It's the paradigm project for road builders.

**Light shines from the** streets far above. Crews have spent over 10 years building the new I-93 tunnel.

Specifically, the Big Dig is the completion of I-90, the longest route on the Interstate, and the reconstruction of I-93, an old elevated highway running through the center of Boston, which is being placed entirely underground. It's the building of 159 lane-miles of highway in a 7.8-mile corridor. When the Big Dig is completed in 2006, its planning and building will have taken over 30 years.

Imagine putting the Panama Canal through Lower Manhattan while keeping the area fully open for business, and you have a good idea of the magnitude and difficulty of the Big Dig. When CBS News and the Travel Channel featured the top 10 transportation projects of all time, they ranked the unfinished Big Dig as number four.

"We are not building a tunnel, we are constructing an underground superhighway," explains Tony Lancellotti, the project's chief engineer. But the Big Dig is not just a tunnel project. In fact, looking at what the Big Dig is not is probably the best way to understand what it is.

It is not an old-style bulldoze-and-pave highway project. Not a single family home is being taken. When it is finished, Boston will be more beautiful than ever, with over 260 acres of new parks, trees, and sidewalks.

It did not stop traffic. Herculean efforts in planning and billions of dollars in the Big Dig's budget have kept I-93 and I-90's traffic running continuously since construction began in 1991. In one of the most densely populated cities in America, the project has kept open major feeder roads, local streets, and even sidewalks. Traffic lights were resynchronized and police details assigned at

**Boston's skyline** is the backdrop to the construction of I-90 and the last few miles of the Interstate System. The Big Dig will pour more concrete than the Hoover Dam. Some of its tunnels have concrete floors 23-feet thick.

every key intersection under construction. Project officials even e-mailed traffic changes to subscribers of the service.

As soon as a new piece of highway was completed, it was turned over to the public for its use.

It's not a barrier to commerce. Business in Boston boomed to record high levels during the busiest years of construction in 1998–2001, burying fears that the construction would kill trade and tourism. The Big Dig has become a tourist attraction itself as visitors pay over $50 for a bike, walking, boat, and/or bus tour of the project.

It's not just a highway project. Boston's mass transit system has been upgraded along the way, and an entirely new subway line was built.

## EXTREME ENGINEERING

One of the first challenges in keeping the existing trains running required under-pinning of the city's 1916 subway tunnel. By 1996, a complex mining operation was under way to create the I-93 northbound tunnel. It was designed to pass beneath the city's Red Line subway, which carries over 200,000 passengers a day. Building the new superhighway tunnel 120 feet below the streets of Boston and within 18 inches of an operating subway system was only one of many challenges.

Project officials called it the largest underpinning of an active subway anywhere in the world. With its 50-foot-high concrete walls and ceiling, it was called the Underground Cathedral by tunnel workers. Either way you saw it, it was a remarkable piece of work.

During construction, Big Dig workers could hear the subways trains rolling over their heads, and if they listened carefully they could hear the warning bells telling passengers that the doors were about to close.

A half mile south of the Underground Cathedral is the east–west highway of I-90. Here, in 1998, workers were preparing to break more records. This time the feat would be described as the largest jack-tunneling and ground-freezing operation ever undertaken.

The problem was that the tunnel to be constructed would run under Boston's South Station, where 80,000 commuter trips begin and end. Keeping these potential drivers off the highways and happy with their trains meant not shutting down or disturbing rail service. However, the tracks the commuters depended on were interconnected. Closing down even one of them to tunnel under it would disrupt the entire system.

Officials, using a concept developed over 100 years earlier in the mining industry, decided to stabilize the

**The Underground Cathedral:** So named by the workers who built this I-93 tunnel beneath a subway tunnel.

ground under the rail lines by freezing it with brine, a salty solution that can reach minus 30 degrees Fahrenheit without solidifying. Running the brine through the earth beneath the rail tracks froze the ground to an ultrasolid state so rail lines wouldn't collapse during construction. After nearly a year of ground freezing, the jacking could begin.

While the ground was chilling, massive tunnel sections, the ones to be jacked—pushed through the frozen ground—were built in large holes immediately next to the tracks. The three tunnels were 167, 258, and 359 feet long, 78 feet wide, and 38 feet high. And each one was heavier than a 16-story building. All of them were outfitted with hefty hydraulic jacks along their back edges. Powered by enormous generators, the jacks drove the tunnel sections forward three feet at a time.

The hulking tunnels stopped when the shields at their leading edge bumped up against the frozen ground ahead of them. From the shield, Big Dig

**The largest jack**
tunneling operation in the world. The tunnels were jacked three feet a day through frozen earth and beneath active trains just five feet above.

workers then dug out a three-foot cavity in front of the tunnel, clearing an underground path for their next push. The routine was repeated for days, months, and years until all three of the new I-90 tunnels completed their odyssey.

Plowing through Boston's transportation history, the tunnels ran up against the foundations of old train stations, seawalls, dock pilings, and granite causeways that carried trains across what was once part of Boston Harbor. When the work was finished, the height, width, and length of all three tunnels made for the largest mass of tunnel ever jacked in the world. After the jacking was complete, it took nearly a year for the frozen earth to thaw.

The list of other engineering feats on the Big Dig is long: the most extensive geotechnical investigation, the widest and first asymmetrical cable-stayed bridge, the largest circular cofferdam and steel-box-girder bridge ever built in the United States, the most extensive use of concrete immersed-tube tunnels in the world, and the first time an active subway has remained operable while an immersed-tube tunnel was built above it. These address only some of the firsts and biggests.

Regardless of the superlatives, every feat was accomplished in the name of keeping travelers on the Interstate System moving through Boston and keeping the city open and prospering. One of the most significant Big Dig legacies will be showing the rest of the country and the world how old roads can be replaced and new ones added while those who depend on them every day continue to use them. A half century after Ike's Grand Plan began, the Big Dig brings it all to an end . . . in a big way.

## THE BIG DIG'S BIG BANG

Over the course of 50 years, the Interstate System has become 42,795 miles of the world's most remarkable bridges, tunnels, and superhighways.

In 1991 the United States Congress and President George Bush approved the Intermodal Surface Transportation Efficiency Act, a 151-billion-dollar, six-year program, bringing to an end the construction of the Interstate System. It also earmarked the last dollars in the Highway Trust Fund to be spent on new Interstate System construction.

Lawmakers singled out a few projects to be the last ones built as part of the original Interstate System. The Big Dig in Boston was by far the largest of these projects, making it the final mile to be built and its dollars the last to be spent in the Grand Plan. All other Interstate System projects from this point forward will be considered improvements to the completed system.

The Big Dig marks the end of the Interstate System's construction; but it also signals the beginning of the next generation of road building, where the project is as much about the communities it passes through as the road itself. Going out in style with the Big Dig creates a Big Bang—an explosion in road-building know-how that creates the standard for the road ahead.

## EPILOGUE

With its 42,795 miles of highway open and carrying millions of Americans to their destinations around the country, the Interstate System envisioned by President Eisenhower and endorsed by the U.S. Congress is complete. But as far-seeing as that original plan was, it didn't go far enough. More Americans are driving more miles to more places than Ike or the Congress could have known in 1956. And so the building and rebuilding of this amazing network continues around the country.

Since 1991, 3882 miles of new highway sporting the red-white-and-blue Interstate shields have been laid around the country, bringing the total miles on the System to 46,677 miles. Additions to the Grand Plan—such as I-69, an 1800-mile-long highway scheduled to pass through eight states and ten cities as it connects Mexico and Canada—are part of the future. Other states have also added miles to the System, such as Pennsylvania's I-99 and Wisconsin's I-39. As long as the Interstate is the highway supporting our society, economy, and national security, it will forever need to be the beneficiary of our attention and investment. The ribbon cuttings will never end!

# Sources

CHAPTER ONE
## THE ROADS THAT BUILT AMERICA
**10 "ribbons across the land":** Dwight D. Eisenhower, *At Ease,* Doubleday & Company, 1967, reprinted by Eastern National, 2000, p. 167.

**11 6000 more people, Nearly two people, Americans put 340 percent:** "The Interstate Highway System," *The Road Information Program (TRIP),* January, 2003, p. 21.

**12 1,000,000-ton pile of:** Angus Kress Gillespie, *Twin Towers,* Penguin Books, Rutgers University Press, NJ, 2002, p. 247.

**13 "Together, the uniting forces":** Dwight D. Eisenhower, White House Press Release, February 22, 1955.

CHAPTER TWO
## AMERICA'S FIRST INTERSTATE
**14 covering 680 miles in five weeks:** Merritt Ierley, *Traveling the National Road,* Overlook Press, Woodstock, New York, 1990, p. 20.

**17 recorded 20 Indian warriors:** Ierley, *Traveling the National Road,* p. 7.

**17 60 men swing, 600 men chopping:** Ierley, *Traveling the National Road,* p. 8.

**17 30 seamen:** Ierley, *Traveling the National Road,* p. 11.

**19 Its right of way:** Norris Schneider, *The National Road: Main Street to America,* The Ohio Historical Society, 1996, p.4.

**19 66-foot-long chain, logarithms:** Karl Raitz, et. al, *The National Road,* The Johns Hopkins University Press, Baltimore & London, 1996, p. 128.

**20 in the spring of 1811:** Raitz, *The National Road,* p. 135.

**21 1000 men at a time, most of them Irish:** Ierley, *Traveling the National Road,* p. 42.

**23 $13,000 per mile:** Schneider, "The National Road," p. 13.

**23 on July 4, 1825:** Schneider, "The National Road," p. 9.

**24 straighter and longer:** Ierely, *Traveling the National Road,* p. 194.

**25 which shall unite by:** Schneider, "The National Road," p. 11.

**26 $6,824,919:** Schneider, "The National Road," p. 15.

**26 1913, Alabama:** *America's Highways 1776-1976,* U.S. Department of Transportation Federal Highway Administration (FHWA), p. 37.

**27 5000 strong:** Ierley, *Traveling the National Road,* p. 62.

**27 shoot his cow:** Schneider, "The National Road," p. 16.

**29 Prussian blue:** Interview: Alan King, National Road Museum, March, 2003.

**29 June Bug, Good Intent, "sober and attentive":** Schneider, "The National Road," p. 22.

**29 eat fresh oysters:** Interview: King, Ohio, March 2003.

**30 "This is the mixing":** Schneider, "The National Road," p. 24.

**31 "water the horses":** Schneider, "The National Road," p. 25.

**33 80-foot-high:** Charles Wingerter, *The History of Greater Wheeling & Vicinity: The Early History of the Wheeling Suspension Bridge,* Chicago, Lewis Publishing, 1912, p. 3.

**33 In 1852:** William Newcott, "America's First Highway," National Geographic, Vol. 193, No.3, March, 1998, p. 93.

**33 The term Parkway:** Witold Rybczynski, *A Clearing in the Distance,* Scribner, New York, 1999, p. 382-383.

CHAPTER THREE
## ROADS OF IRON AND MUD
**35 The spike and hammer:** Stephen Ambrose, *Nothing like It in the World,* Simon & Schuster, New York, 2000, p. 364.

**35 the shot at Concord and Lexington, liberty bell:** Ambrose, "Nothing like It," p. 366.

**39 $10,000:** Richard Weingroff, "A Partnership That Makes a Difference," FHWA, 2002, p. 5.

**39 Fifth Avenue:** Jill Hecht, "Where the Rubber Meets the Road," *Technology Review,* Cambridge, Massachusetts, Nov-Dec, 1998.

**40 The first section of test road:** *America's Highways 1776-1976,* p. 45.

**41 "An itinerant college on wheels," "This college does not," $80,000, October 29, 1901, 4037 miles:** *America's Highways 1776-1976,* p. 50.

**41 Teddy Roosevelt:** www.angelfire.com, June 2003.

**42 $50:** Stephen Sears, "Ocean to Ocean in an Automobile Car," *American Heritage,* 1980, pp. 58-64.

**43 1884:** bicyclemuseum.com, February, 2002.

**46 $8000 in wear and tear:** Sears, "Ocean to Ocean in an Automobile Car," pp. 58-64.

**46 "Dr. H. Nelson":** Ralph Nading Hill, *The Mad Doctors Drive,* Stephen Greene Press, Brattelboro, Vermont 1964, p. 33.

**48 2,151,570:** Weingroff, "Partnership," p. 13.

**49 8000 trucks:** Paul Dickson and William Hickman, *Firestone, A Legend, A Century, A Celebration,* Forbes Custom Publishing, New York, NY, 2000.

**49 Harvard College:** Bruce Seely, *Building the American Highway System,* Temple University Press, Philadelphia, 1987, p. 13.

**50 "Ship by truck":** Dickson and Hickman, *Firestone.*

**51 April 9, 1917:** Jeffrey L. Rodengen, *The Legend of Goodyear, The First 100 Years,* Write Stuff Syndicate, Inc., Fort Lauderdale, FL, 1997, p. 48.

**51 the run of 21 days:** Rodengen, "The Legend of Goodyear," p. 49.

**51 exchanged for new, 3700-mile trip, first-ever sleeping:** Rodengen, "The Legend of Goodyear."

**51 30,000 trucks:** *America's Highways 1776-1976,* p. 95.

**52 On November 22:** *America's Highways 1776-1976,* p. 95.

CHAPTER FOUR
## THE GREAT AMERICAN PARTNERSHIP
**56 "a lot of human vultures":** Weingroff, "Partnership," pp. 11-15.

**57 government predicted:** *America's Highways 1776-1976,* p. 46.

**58 In 1893, Massachusetts created the nation's first:** *America's Highways 1776-1976,* p. 43, p. 65.

**59 California was the first:** Weingroff, "Partnership," p. 45.

**59 1870-1918:** Weingroff, "Partnership," p. 48.

**61 "Those who do not have":** Seely, *Building the American Highway System,* p. 76.

**61 40,000,000 pounds of, $10,000,000 pile:** *America's Highways 1776-1976,* p. 105.

**63 $75,000,000:** Tom Lewis, *Divided Highways,* The Penguin Group, New York, 1997, p. 18.

**63 In 1922, MacDonald:** Weingroff, "Partnership," p. 65.

**63 "My aim is this":** Lewis, *Divided Highways,* p. 10.

**63 The Revenue Act of 1932:** James Sweet, *The Federal Gas Tax At A Glance,* Bybee House, 1993, p. 5.

CHAPTER FIVE
## A RADICAL IDEA: NUMBERED HIGHWAYS

**64 By 1916:** Weingroff, "Partnership," p. 25.

**66 There were over 50 interstate:** *America's Highways 1776-1976,* pp. 109-110.

**67 and 21 state highway:** *America's Highways 1776-1976,* p. 110.

**68-69 Big Road Advocate:** AAA Press-Materials, 2003.

**70 81,096 miles:** *America's Highways 1776-1976,* p.110.

**70 Oregon was the first:** Sweet, *The Federal Gas Tax At A Glance,* p.5.

**71 The five chief engineers:** Jim Powell, "The U.S. Numbered Highway System," *Show Me Route 66 - 75th Anniversary,* Volume 12, Number 4, November, 2001, p. 19.

**72 20,000:** *America's Highways 1776-1976,* pp. 127-128.

**73 145 U.S. routes:** Powell, "The U.S. Numbered Highway System," p. 20.

**74 132 legitimate:** Powell, "The U.S. Numbered Highway System," p. 20.

**75 his green 1941:** Michael Wallis, *Route 66, The Mother Road,* St. Martin's Press, New York, NY, 2001, pp. 9-15.

**76 who spent 116 TV Episodes:** Jim Powell, "Route 66 Timeline," *Show Me Route 66 - 75th Anniversary,* Volume 12, Number 4, November, 2001, p. 11.

**76 whispered to him, "Route 66 changed":** Wallis, "Mother Road," Back cover.

**77 Route 66 first aired:** Wallis, "Mother Road," p. 268.

**77 On January 1977:** Wallis, "Mother Road," p. 29.

**77 1162 miles:** Jim Powell, "Route 66 Association of Missouri," *Show Me Route 66 - 75th Anniversary,* Volume 12, Number 4, November, 2001, p. 2.

**77 "The death":** Interview: Weingroff, June, 2002.

**77 "We are finding that":** Powell, "Route 66 Timeline," p. 13.

**78 "The traveler may shed, if numbering our":** Weingroff, "Partnership," p. 63.

**78 an army of up to 480,000, than ever before-$1,000,000,000:** *America's Highways 1776-1976,* p. 125.

**81 "I have just been fired":** Lewis, *Divided Highways,* p. 92.

**82 "I want these":** Dan Cupper, *The Pennsylvania Turnpike History,* Applied Arts Publisher, Lebanon, Pennsylvania, 1995, p. 9.

**82 939 local roads, 307 bridges:** Cupper, *The Pennsylvania Turnpike History,* p. 11.

**82 10,000 cars a day:** Cupper, *The Pennsylvania Turnpike History,* p. 22.

CHAPTER SIX
## IKE'S GRAND PLAN

**85 moving more soldiers, firepower:** Stephen Ambrose, *Citizen Soldiers,* Simon & Schuster, New York, NY, 1997, p. 201.

**86 "As the old expression went":** Eisenhower, *At Ease,* p. 70.

**86 "Paving was unknown":** Eisenhower, *At Ease,* p. 68.

**87 1897, 16-gauge:** Eisenhower, *At Ease,* p. 108.

**87 "A feeling came over me":** Eisenhower, *At Ease,* pp. 3-6.

**87 61 out of 164:** www.eisenhower.utexas.edu, October, 2002.

**87 The English translation:** Eisenhower, *At Ease,* p. 56.

**88 Asiatic enemy, 350 cities, over 3,250,000 Americans cheered:** Captain William Greany, Adjunct and Statistical Officer, "Principal Facts Concerning the First Transcontinental Army Motor Transport Expedition, Washington to San Francisco," U.S. Army Memorandum, July 7 to September 6, 1919.

**88 "Proceed by way of the Lincoln Highway to San Francisco without delay":** Weingroff, *Public Roads,* March/April 2002, p. 22.

**93 stretched out over three miles:** Lewis, *Divided Highways,* p. 89.

**94 "The trip had been difficult":** Eisenhower, *At Ease,* pp. 166-167.

**94 "No sooner had we left":** Eisenhower, *At Ease,* p. 167.

**94 against 275:** Stephen Ambrose, *Eisenhower: Soldier and President,* Touchstone, New York, NY, 1991, p. 41.

**94 French highways:** Ambrose, *Eisenhower: Soldier and President,* p. 43.

**94 "This was the greatest disappointment":** Eisenhower, *At Ease,* p. 181

**95 "You will land in":** "Eisenhower," National Park Service, U.S. Department of the Interior, 1992.

**95 "The most difficult":** Stephen Ambrose, *D-Day,* Touchstone, New York, NY, 1995, p. 13.

**97 76,000 tons of:** Ambrose, *D-Day,* p. 97.

**97 1200 army truck drivers, nearly 80 percent of:** Ambrose, *Citizen Soldiers,* p. 372.

**97 Between August 29, 6,000 trucks:** Ambrose, *Citizen Soldiers,* p. 113.

**98 "Hitler's Weather":** Edward Oxford, "Our Greatest Land Battle," *American History,* Feb/Mar, 1995, p. 5.

**98 11,000 men, 250,000 soldiers:** Ambrose, *Citizen Soldiers,* p. 201.

**98 Senator Albert Gore, Sr.:** Associated Press "Albert Gore Sr. recalls fight for interstate highway bill," *Bristol Herald Courier,* June 16, 1996.

**99 2400 miles:** Richard Weingroff, "The Man Who Changed America," *Public Roads,* Draft Copy, December, 2002, p. 7.

**100 16,000 of them a day:** Lewis, *Divided Highways,* p. 104.

**102 suicide alleys:** Author's Interview: Dan McNichol, author's father, January 2003.

**102 In 1949 the 37,681 miles of highway slated:** *America's Highways,* p. 158.

**103 "Our highway net," July 12, 1954:** Weingroff, " The Man Who Changed America," p. 14, p. 29.

**103 killing 40,000 people:** Weingroff, " The Man Who Changed America,"p. 15.

**104 70,000,000 urban residents:** Lewis, *Divided Highways,* p. 108.

**105 292 to 123:** Weingroff, "The Man Who Changed America," p. 22.

**105 In 1953:** Lewis, *Divided Highways,* p. 107.

**106 388 to 19:** Weingroff, "The Man Who Changed America," p. 4.

**106 understated that Ike was "highly pleased":** Weingroff, "The Man Who Changed America," p. 29.

**106 100,000 people:** Lewis, *Divided Highways,* p. 209.

**107 "I see an America":** Campaign Speech, Lexington Kentucky, October 1, 1956.

**107 1956 with the highest:** "Eisenhower," National Park Service, U.S. Dept. of the Ineterior, G.P.O., 1993.

**107 More than any single action:** Weingroff, "The Man Who Changed America," p. 38.

**108 "Levittown," "it was universally,":** Lewis, *Divided Highways,* p. 77.

**109 With 82,000, 17,400:** Gwendolyn Wright, *Building the Dream,* Pantheon Books, New York, 1981, p. 252.

**109 27 clearly defined steps:** Kenneth Jackson, *Crabgrass Frontier: The Suburbanization of the United States,* Oxford University Press, New York, 1985, p. 231-238.

CHAPTER SEVEN
## THE INTERSTATE DECADE

**112 "The Interstate Decade":** William Haycraft, *Yellow Steel,* University of Illinois Press, Chicago, 2002, p. 143.

**112 "This Nation doesn't have":** Robert Paul Jordan, "Our Growing Interstate Highway System," *National Geographic,* p. 198.

**114 1,600,000 acres of land, 1972:** Jordan, "Growing Interstate," p. 198.

**114 48,000 workers:** "The Interstates and the National Economy," *Go Magazine,* August, 1966.

**115 $27,000,000, 836 separate sections, 24-ton:** *American Highways,* pp. 334–335.

**115 On August 13, 1956:** Richard Weingroff, *Public Roads Special Addition,* 1996, p. 17.

**116 16-bridge, scaled-down:** *American Highways,* p. 335.

**117 They had survived 1,114,000:** *America's Highways,* pp. 334–335.

**118 $200,000,000:** Jordan, "Growing Interstate," p. 206.

**119 metropolis, utopia, green for every one that sided:** Lewis, *Divided Highways,* p. 137.

**123 2,102 miles:** Haycraft, *Yellow Steel,* p. 144.

**125 using nuclear bombs:** Jim Drago, "Amazing But True," *Going Places,* March/April 1988, p. 13-17.

**126 "I used to tell our people":** The American Association of State Highway and Transportation Officials, "The States and the Interstates," Washington, D.C., 1991, p. 51.

**127 In 1859, The Suez canal:** Haycraft, *Yellow Steel,* p. 7.

**127 In 1904 Americans took over, 262 million:** Haycraft, *Yellow Steel,* p. 9.

**127 42 billion cubic:** Interview: Haycraft, author: *Yellow Steel,* January 14, 2003.

**128 "Mother, Father," "come home," "It's almost,":** Robert B. Thompson, "Lust See TV," *Washington Post Magazine,* pp. 16-17.

**128 "Miss Cover," "Now, everyone,":** "First Women Highway Engineer in Public Roads," *The News In Public Roads,* United States Department of Commerce, Bureau of Public Roads, April, 1962, p. 1.

**129 "I am not allowed" 'Get those women," "$8 to $10":** Weingroff, "40th Anniversary of Dwight D. Eisenhower System of Interstate and Defense Highways / I-70," FHWA, 1998, p. 7.

**129 "They had a woman":** Lewis, *Divided Highways,* p. 257.

**129 "How proud Ike":** Weingroff, "40th Anniversary...," pp. 6-7.

**130 Hi-Way-Yellow:** Haycraft, *Yellow Steel,* p. 64.

**130 a wall nine feet thick:** Robert, "Our Growing Interstate Highway System," p. 198.

**131 largest machines ever built:** Interview: Jerry Shananhan, Caterpillar Dealership, February, 2003.

**132 "He was good enough":** Clem Work, "Chip Dinosaur out of River Bed," *The Rocky Mountain News,* October 11, 1970.

**134 The Interstate System consumed:** John Gibbons, "The Incredible Interstate System," *Highway Users Quarterly,* Spring 1975, p. 15.

**136 $17,000:** *America's Highways,* p. 113.

**136 600 feet an hour:** Interview: Jim Rodriguez, CMI Vice President, June, 2003.

**137 5,201 miles:** Interview: Weingroff, January, 2003.

CHAPTER EIGHT
## THE INTERSTATE GOES DOWNTOWN

**139 against his wishes:** General Bragdon's White House Memorandum for the Record: April 8, 1960.

**142 "the hub of the solar system":** Thomas O'Connor, *Boston A-Z,* Harvard University Press, Cambridge Massachusetts, 2001, p. 169.

**143 "I only wish":** Yanni Tsipis, *Building the Mass Pike,* Arcadia Publishing, Charleston, South Carolina, 2002, p. 49.

**144 December 16, 1773:** O'Connor, *Boston A-Z,* p. 51.

**146 $18,000,000 price:** Interview: Chan Rogers, engineer, March 2003.

**148 "Callahan's Folley," "This new highway,":** Yanni Tsipis, Draft: *Building Route 128.*

**149 "Cabot, Cabot and Grab-It,"**: Tony Hill, "Route 128, The Baby Boomer of highways, turns 50," *The Boston Globe*, August 19, 2001.

**151 "Never, Never"**: Alan Lupo, et, al, *Rites of Way*, Little Brown, Boston, 1971, p. 47.

**154 "When we were looking for"**: Interview: Rogers, March 2003.

**155 1965 Watts**: David Brodsly, *L.A. Freeway*, University of California Press, Los Angeles, 1981, p. 35.

**155 Romney**: Helen Leavitt, *Superhighway - Superhoax*, Doubleday & Company, Inc. Garden City, New Jersey, 1970. p. 6.

**155 "rather grotesque," "If we had to do it again"**: Lewis, *Divided Highways*, p. 188.

**155 the nation's longest row**: Lewis, *Divided Highways*, pp. 210-211.

**156 "You got to"**: Interview: Pete Mainville, December 2002.

**157 "I jumped up on a table"**: Senator Barbara Mikulski, et. al, *Nine and Counting: The Women of the Senate*, Harper Collins, New York, 2000, pp. 30-31.

**158 400 black**: Mikulski, "Nine and Counting," pp. 30-31.

**158 "My name is Frank"**: Mikulski, "Nine and Counting," pp. 30-31.

**159 342-acre**: Interview: Bill Bearden, park advocate, March, 2003.

**159 "little old ladies in tennis shoes"** Interview: Bearden, April 2003.

**159 July 30, 1954**: Interview: Bearden, March, 2003.

**159 10,000 petitions**: www.kurumi.com, July, 2003.

**160 842 homes, 44 businesses**: Interview: Bearden, March 2003.

**160 On January 26, 1981**: www.kurumi.com, July, 2003.

**160 "The assassination"**: Interview: Bearden, March, 2003.

## CHAPTER NINE
## 54,633 BRIDGES

**163 54,633 Bridges**: FHWA Interstate Bridge Table, December, 2002.

**164 Nearly 25%**: "Highway Infrastructure," *GAO Report to the Chairman, Committee on Transportation and Infrastructure, House of Representatives*, United States General Accounting Office, May, 2002, p.48.

**166 the world's largest American Flag**: www.nycroads.com, May, 2003.

**167 300,000 vehicles**: ibid.

**168 10,176-foot**: www.dot.ca.gov, May, 2003.

**172 including twice in a day**: ibid.

**174 world's record for the heaviest lift**: www.travelportland.com, May 2003.

**175 9/16 of an inch thick**: Correspondence via e-mail: Patrick Martens (MoDOT) May, 2003.

**176 5,400, 6,230**: www.opacengineers.com, May, 2003.

**181 "Most magicians are"**: Barb Kampbell, "Top of the News," *The Trucker*, August 23, 2002.

**183 killing 77 people**: www.chron.com, May, 2003.

**184 87,000-ton vessel**: *Civil Engineering*, November/December 2003, p. 163.

## CHAPTER TEN
## 104 TUNNELS

**193 Title of "104 tunnels"**: Interview: Anthony Caserta, FHWA tunnel engineer, April 2003.

**194 seven years, 13 workers, 40 feet**: www.nycroads.com, January, 2003.

**195 original 84 fans**: Anthony Caserta, FHWA written comments, June 2003.

**195 On November 12, President Calvin Coolidge**: Albin Krebs, "Ole Singstad, 87, Master Builder of Underwater Tunnels," *The New York Times*, 1969.

**196-197 11,155 feet above sea level, nine men would die, beneath a ski resort, tourism has boomed**: Merritt, George, "Conquering the Divide," *The Denver Post*, Associated Press, March 9, 2003.

**196 1867 W.A. Loveland looks for a way to run tracks through mountain**: Merritt, "Conquering the Divide."

**196 200,000,000 vehicles have passed**: TheDenverChannel.com, October 25, 2001.

**197 "We were going by the book"**: www.dot.state.co.us/Eisenhower, December 27, 2002.

**197 1,140 tunnel workers**: Kevin Flynn, "McOllough Engineered Eisenhower Tunnel," *Rocky Mountain News*, January 16, 2003.

**199 "The largest single"**: Brochure, Construction of I-95 Fort McHenry Tunnel, Interstate Division of Baltimore, 1983, p. 1.

**199 each weighing 31,882 tons**: "Construction of the I-95 Fort McHenry Tunnel."

**199 56-mile, 32 of these sections, 350 feet**: Corinne Bernstein, *Civil Engineering*, July 1986, p. 38.

**200 The U.S. Military**: Interview: Tony Lancellotti, Big Dig engineer, May, 2003.

**201 "a project that just made"**: Interview: Ken Merrill, construction manger, April 2003.

**213 $38,000,000**: Anthony Caserta, FHWA written comments, June 2003.

**215 700 feet**: Interview: William Connell, Big Dig ventilation expert, May 2003.

**216 2,400 degrees**: Interview: Sergiu Luchian, Big Dig engineer, May 2003.

**216 The highest elevation**: Weingroff, 40th Anniversary of Dwight D. Eisenhower System of Interstate and Defense Highways/ I-70, May 1, 1998.

## CHAPTER 11
## THE ROAD AHEAD

**219 "The Interstates Affect"**: Interview: Weingroff, January 2002.

**220 "We're recruiting at the prison gates"**: "Boom is Riding South Along Interstate 85," *The New York Times*, July 8, 1966.

221 **"One thing that's happened is the color line,"** **"Economics has brought":** "Boom is Riding South Along Interstate 85."

221 **$400,000,000 assembly, 10 nations, $36 million, Autobahn:** Cover Story: "The Boom Belt," *Business Week,* September 27, 1993, pp. 98-154.

221 **140 homeowners:** "The Boom Belt," pp. 98-154.

222 **Howard Dearing Johnson borrowed $2,000:** www.hojo.com, May 2003.

223 **"Those still around...Burger Boy Food-O-Ramas":** Eric Schlosser, *Fast Food Nation,* Houghton Mifflin Company, New York, 2002, p. 18-23.

226 **only five percent:** Jerry Adler, "Bye, Bye Suburban Dream," *Newsweek,* May 15, 1995, p. 49.

226 **1.2 million acres of:** John G. Mitchell, "The American Dream-Urban Sprawl," *National Geographic,* July 2001, p. 58.

226 **20 miles:** Interview: Weingroff, July, 2003.

225 **By the mid 1970s, by the 1980s:** Mitchell, "The American Dream," p. 64.

227 **42 percent:** "The Interstate Highway System: Saving Lives, Time and Money, But Increasing Congestion Threatens Benefits," Washington, D.C., The Road Information Program (TRIP), January 2003, p. 4.

227 **Over one million more trucks:** Joel Achenbach, "Rough Draft: Dealing with the 18-wheelers," *The Washington Post,* July 8, 2000.

227 **Ninety-three percent:** "Our Nation's Highways-Selected Facts and Figures," USDOT/FHWA, Washington, D.C., 1996, p. 4.

228-229 **An Interstate Success Story: UPS:** UPS Press-Materials, 2003.

228 **"At one point or another":** Interview: Dan McMackin, UPS spokesman, May 2003.

229 **"When 95 wasn't":** Interview: Dick Socha, driver, May 2003.

230 **66 percent, $23 billion, $125 billion:** "The Interstate Highway System," pp. 17, 26-27.

230 **There are 1,000,000:** Achenbach, "Rough Draft," *Washington Post,* July 8, 2000.

231 **American Trucking Association estimates that 65 percent:** Peter Kilborn, "Interstates Give Boost to Rural Economics," *The New York Times,* July 14, 2001, p.1.

232 **"Old, grouchy highway":** *The Builders,* National Geographic Book Division, National Geographic Society, Washington, D.C.

232 **$35, $45, $22,000:** "Glenwood Canyon, The Final Link," United States Department of Transportation, 1992, p. 31.

232 **bighorn sheep, 150,000 plants:** "Glenwood Canyon, The Final Link," p. 5.

234 **3900-foot-long, only major tunnels able to support themselves:** *Technology Review (TR) News,* March/April, 1991, p. 33.

234 **accommodate the military's need:** *American Highways,* April 1960.

235 **$50,000,000:** Jill Hecht, *Technology Review,* Cambridge, Massachusetts, November/December, 1998.

236 **$100,000 an hour:** Interview: Mickey McGee, Georgia DOT, May 23, 2003.

237 **"It would not," 32 lane-miles, 540 days, 22 weekends, as many as 14:** "Asphalt Reconstruction of the Atlanta Beltway," (Cover Article) *Public Works,* January, 2002.

238 **a million miles, one accident:** Interview: Colin Ahsmore, Westrack Project Manger, May 21, 2003.

240 **1.8-mile track, 26 sections:** Interview: Colin Ahsmore, May 21, 2003.

241 **"We are not building a tunnel":** Interview: Lancellotti, April, 2000.

243 **167, 258, and 359 feet long:** White Paper-Central Artery/Tunnel Project Document with Table.

245 **3,882 miles:** Interview: Weingroff, June 2003.

# Index

Page numbers in **bold** refer to illustrations and captions.

Publisher: Barbara J. Morgan
Editor: Marjorie Palmer
Production: Della R. Mancuso
Mancuso Associates, Inc.,
North Salem, NY

# Bibliography

The following books, articles, and interviews were instrumental in the writing of this book:

BOOKS:

Ambrose, Stephen, *Citizen Soldiers,* New York, Simon & Schuster, 1997; ———, *D-Day,* New York, Touchstone, 1995; ———, *Eisenhower: Soldier and President,* New York, Touchstone, 1991; ———, *Nothing Like It in the World, 1863–1869,* New York, Simon & Schuster, 2000; American Association of State Highway and Transportation Officials, *Interstate: The States and the Interstates,* Washington, D.C., 1991; Clymer, Floyd, *Treasury of Early American Automobiles: 1877–1925,* New York, McGraw-Hill, 1950; Colley, David, *The Road to Victory: The Untold Story of WWII's Red Ball Express,* Washington, D.C., Brassey's, 2000; Cupper, Dan, *The Pennsylvania Turnpike,* Lebanon, Pa., Applied Arts, 1995; Eisenhower, Dwight D., *At Ease: Stories I Tell Friends,* New York, Doubleday, 1967; Federal Highway Administration (FHWA), Department of Transportation, *America's Highways: 1776/1976,* Washington, D.C., U.S. Government Printing Office, 1976; Freeth, Nick, *Route 66,* St. Paul, MBI, 2001; Haycraft, William R., *Yellow Steel,* Chicago, University of Illinois Press, 2002; Ierley, Merritt, *Traveling the National Road,* New York, Overlook Press, 1990; Jackson, Kenneth, *Crabgrass Frontier: The Suburbanization of the United States,* New York, Oxford University Press, 1985; Kay, Jane Holtz, *Asphalt Nation,* Los Angeles, University of California Press, 1997; Koolhaas, Rem, et al., *The Harvard Design School Guide to Shopping,* New York, Taschen-America, LLC., 2001; Leavitt, Helen, *Superhighway —Superhoax,* New York, Doubleday, 1970; Lewis, Tom, *Divided Highways,* New York, The Penguin Group, 1997; Lupo, Alan, *Rites of Way,* Boston, Little, Brown, 1971; Mikulski, Senator Barbara, et al, *Nine and Counting: the Women of the Senate,* New York, HarperCollins, 2000; Raitz, Karl, et al., *The National Road,* Baltimore, Johns Hopkins University Press, 1996; Rose, Mark, *Interstate: Express Highway Politics 1939–1989,* Knoxville, University of Tennessee Press, 1990; Schneider, Norris, *The National Road, Main Street of America,* Columbus, The Ohio Historical Society, 1996; Seely, Bruce, *Building the American Highway System,* Philadelphia, Temple University Press, 1987; Tobin, James, *Great Projects,* New York, The Free Press, 2001; Tsipis, Yanni, *Route 128,* Charleston, Arcadia, Draft Copy, 2003;

ARTICLES:

Adler, Jerry, "Bye, Bye Suburban Dream," *Newsweek,* May 15, 1995; "Asphalt Reconstruction of the Atlanta Beltway," *Public Works,* January 2002; "The Boom Belt," *Business Week,* September 27, 1993; "Boom Is Riding South Along Interstate 85," *The New York Times,* July 8, 1966; Civil Engineering, November/December 2002; *The Dwight D. Eisenhower System of Interstate and Defense Highways, Route Log Finder List,* FHWA, November 2002; Egan, Timothy, "Urban Sprawl Strains Western States," *The New York Times,* December 29, 1996; "Highway Infrastructure," *GAO Report to the Chairman, Committee on Transportation and Infrastructure,* House of Representatives, United States General Accounting Office, May 2002; "The Interstate Highway System: Saving Lives, Time and Money, But Increasing Congestion Threatens Benefits," The Road Information Program, TRIP, January 2003; Jordan, Robert Paul, "Our Growing Interstate Highway System," *National Geographic,* pp.194–219, February 1968; Merritt, George, "Conquering the Divide," *The Denver Post,* March 9, 2003; Mitchell, John G., "The American Dream—Urban Sprawl," *National Geographic,* July 2001; Powell, Jim, *Show Me Route 66—75th Anniversary,* Vol. XII, No. 4, November 2001, Route 66 Association, Inc., 2001; Satchell, Michael, "Trials in the Tunnel of Terror," *U.S. News & World Report,* December 31, 2001/ January 7, 2002; Schonfeld, Erick, "The Total Package," *e-Company Now,* June 2001; Sears, Stephen W., "Ocean to Ocean in an Automobile Car," *American Heritage,* pp. 58–64, 1980; Weingroff, Richard, "A Partnership That Makes a Difference: An Anniversary Look at 1916 and 1956," 2002; ———, *Public Roads: Special Addition 1996,* FHWA, 1996; ———, "The Man Who Changed America," *Public Roads,* FHWA, Draft Copy, December 2002.

INTERVIEWS:

Ahsmore, Colin, Westrack Project Manager, May 2003; Matthews, Dr. James K., Director, United States Transportation Command, March 2003; Salvucci, Fred, Professor, Massachusetts Institute of Technology, January 2002, March 2003; Smith, Don, Caterpillar Corporation, May 2003; Valle, Jan, Escort for Caterpillar's 345, May 2003; Yasui, Glenn, Hawaii Department of Transportation, September 2002.

# Acknowledgments

I am forever grateful to my publisher, Barbara Morgan. She changed my life three years ago, providing me the opportunity of a lifetime when she asked me to pen the first book on the Big Dig. Without the vision, beliefs, and patience of Barbara and of my editor, Marjorie Palmer, this book wouldn't have been possible. The sage advice and graphic talents of Richard Berenson make the story a seamless visual trip. I have come to depend on his judgment and friendship. Jane Neighbors's superb copyediting skills are evident on every page. The editorial team is complete with Emily Seese, who coordinated the seemingly endless logistics—a million thanks for her intellectual curiosity and support.

Richard Weingroff, the Federal Highway Administration's sole historian, is a remarkable man, not only for his depth of knowledge about America's highways but for his determination in keeping the record straight and his patience for all of us who are trying to grasp what he knows. He was central to the writing of this book and his contributions cannot be quantified. I owe him and everyone at the FHWA a heavy debt of gratitude.

Professor Gretchen Schneider guided me through the history and academic predictions for the American cities and their suburbs. She wrote the sidebars on retail businesses and Levittown and selflessly contributed her time, opinions, and remarkable energies to this book. Alfred Chock, engineer and writer, created the text for chapter nine and assisted me with its subjects and the bridge-disaster sidebar. He gave generously of his time. Jim Powell, founder of the Missouri Route 66 Association also gave his time generously and made critical reviews of the text. Melissa Nicienski's extraordinary skills at research provided voluminous amounts of material, no matter how arcane the request. Yanni Tsipis lent his talents as an engineer and historian, especially with the sidebar, America's First Beltway. He led me to many of the images in chapter eight.

Caterpillar Inc. generously donated a number of the images in the book, and Alex Campbell made the company's many resources available to me. Don R. Smith of Caterpillar pieced together the accounts of September 11, 2001, and Elmer and Shirley Hershberger and Jan Vallee provided extraordinary accounts. Bill Haycraft shared his deep knowledge of Caterpillar and other heavy equipment companies. Jeff Wales, Ben Thirkield, Jerry Shananhan, and Peter D'Agostino of Caterpillar's Southworth-Milton dealership shared their world of heavy iron with me. UPS's Dan McMackin invited me to "Brown up" and ride with Richard Socha, who shared his experiences in over three decades of Interstate trucking.

The following people reviewed the manuscript and/or made invaluable contributions to it: My father Dan McNichol, Bernie McCabe, Joe Sharkey, and Chan Rogers of the construction world; Tony Caserta, Terry Mitchell, Abe Wong, and Dan Wood of FHWA, and its current and former administrators Mary Peters and Thomas Larson. James Matthews explained the military uses of the Interstate during mobilizations. Tony Lancellotti, Christopher Reseigh, Pete Mainville, Bill Connell, David Sailors, Brian Brenner of Parsons Brinckerhoff; many of the staff and alumni of Wentworth Institute of Technology, as well as professors Fred Salvucci and Ken Krukemeyer of MIT steered me through historic and technical channels. David Newcomb, Colin Ashmore, and Buzz Powell contributed to the explanation of all things asphalt.

I am grateful to: Matthew Amorello, Jack Quinlan, Sean O'Neill, Danielle Bowman, Phil De Joseph, Dennis Rahilly, and Sergiu Luchian of the Big Dig; state officials: Dennis Trujillo of Caltrans, Keith Duerling of Maryland Transportation Authority, Mickey McGee, of Georgia DOT, Alan King of the Ohio's National Road Museum, and librarian Lynn Matis. A special thanks to sounding boards: Laura Moloney, Megan Longley, Greg Wolfe and Chris Nagle. Finally, deepest appreciation to my mother and father and my fiancée Jin Ji for their endless love and support over the last two years.

# Picture Credits

Pages 8-9 © Alex S. MacLean/Landslides; 10, 11 © Dan McNichol; 13 courtesy Caterpillar, Inc.; 14 Illustration by Howard Pyle. *Harper's Weekly,* March 12, 1881. Collection of the author; 16 The George Washington Papers at the Library of Congress; 17 © CORBIS; 18 left © Bettmann/CORBIS; 18 right © CORBIS; 20 bottom National Archives; 20–21 top Illustration in *Traveling the National Road,* New York, Overlook Press, 1990; 21 bottom After John T. Hriblan (HTIA); 23 U.S. Bureau of Public Roads/National Archives; 24 Illustration in *Harper's Magazine,* November 1879; 25 Carl Rakeman, courtesy Texas Transportation Institute; 27 Illustration in *Harper's Magazine,* November 1879 (left); 27 top © David H. Wells/CORBIS; 27 bottom © David H. Wells/CORBIS; 28 © Bettman/CORBIS; 29 Library of Congress; 30–31, 32–33 The Brown Collection of Photographs, 1856. Service provided by the staff of the Ohio County Public Library in partnership with and funded in part by the Wheeling National Heritage Area Corporation. 34-35 © CORBIS; 35 right © Dan McNichol; 37 left © Bettmann/CORBIS; 37 right Library of Congress; 38 © Bettmann/CORBIS; 40 United States Department of Transportation-Federal Highway Administration (USDOT/FHWA); 41 USDOT/FHWA; 42 Division of Transportation, Smithsonian Institution; 43, 44, 45 Wilbur Collection, University of Vermont Library, Burlington; 47 ©PEMCO-Webster & Stevens Collection; Museum of History & Industry/CORBIS; 48 Automobile Manufacturer's Association, Inc.; 50 The Goodyear Tire & Rubber Co.; 51 Chicago Historical Society. 1917; 52, 53 USDOT/ FWHA; 55 ©PEMCO-Webster & Stevens Collection; Museum of History & Industry, Seattle/CORBIS; 56 USDOT/FWHA; 57 top© Denver Public Library/CORBIS; 57 bottom USDOT/FHWA; 58 US Bureau of Public Roads (National Archives); 60, 61, 62, 63 USDOT/FHWA; 64 Carl Rakeman; courtesy Texas Transportation Institute; 66 Library of Congress; 67, 68, 69 Courtesy Automobile Association of America (AAA); 70 © Jim Powell; 72, 73 USDOT/FHWA; 74 Library of Congress; 75 top Library of Congress, 75 bottom © CORBIS; 76 © Lake County Museum/CORBIS; 75 bottom © Rykoff Collection/CORBIS; 77 © Bettmann/CORBIS; 78 Library of Congress; 79 courtesy Ohio Department of Transportation; 80 Carl Rakeman, courtesy Texas Transportation Institute; 82 © CORBIS; 83 top and bottom © CORBIS; 83 middle USDOT/FHA; 84, 85 © Hulton-Deutsch Collection/CORBIS; 86, 87 Dwight D. Eisenhower Library, Abilene, KS; 88 Library of Congress; 89 top Library of Congress; 89 bottom National Archives; 90–91 Dwight D. Eisenhower Library, Abilene, KS; 92 US Signal Corps/National Archive; 94 Dwight D. Eisenhower Library, Abilene, KS; 95 top © Bettman/CORBIS. June 1944; 95 bottom © CORBIS; 96, 97 © CORBIS; 99 © Bettman/CORBIS; 100–101 © CORBIS; 102, 104 © Bettmann/CORBIS; 105 USDOT/FHWA; 106 © Bettmann/CORBIS; 107 USDOT/FHWA; 108-109, 110-111 © Bettmann/ CORBIS; 112-113 Courtesy Caterpillar, Inc.; 115, 116–117, 118, 119, 120, 121, 122 USDOT/FHWA; 124 *LIFE* Magazine, June 4, 1956; 129 USDOT/FHWA; 131 USDOT/ FHWA; 132 Reprinted with permission of *The Rocky Mountain News;* 134–135 USDOT/FHWA; 136 courtesy CMI Terex Corporation; 137 courtesy Maryland Transportation Authority; 138–139, 140 USDOT/ FHA; 142 collection of the author; 143 top courtesy The Barletta Corporation; 143 left courtesy Perini Corporation; 143 right collection of the author; 144, 145 courtesy Massachusetts State Transportation Library; 146, 147 © Dan McNichol; 149 left courtesy Massachusetts State Archives; 149 right courtesy Perini Corporation; 150, 151 courtesy Massachusetts State Transportation Library; 152 top Central Artery/Tunnel Project; 152 middle Massachusetts Highway Department; 152 bottom © Dan McNichol; 153 © Dan McNichol; 154 courtesy Chan Rogers; 155 Louisiana Photograph Collection. Municipal Government Collection; Department of Streets Series; 156-157 © Michael Higgins; 158 top © Kevin Fleming/CORBIS; 158 bottom ©AFP/CORBIS; 159 © Mapquest.com; 160, 161 The William Bearden Co., Inc.; 162–3 © Richard A. Cooke/CORBIS; 165 courtesy CALTRANS; 166 © Bob Krist/CORBIS; 167 left Library of Congress HABS/HAER; 167 right courtesy The Port Authority of New York and New Jersey; 168 © CORBIS; 169 courtesy Caterpillar, Inc.; 170 © Kit Kittle/CORBIS; 171 © Bettmann/CORBIS; 172–173 courtesy CALTRANS. Photographer Bob Colin; 176 © Dan McNichol; 177 © David Sailors; 178, 179 © Bettmann/ CORBIS; 180 USDOT/FHWA; 181 ©Reuters NewMedia Inc./CORBIS; 182–183 Michael Higgins; 184 © Joseph Sohm/CORBIS; 185 © Raymond Gehman/CORBIS; 186–187 © Dan McNichol; 189 © Paul A. Souders/CORBIS; 191 © David Sailors; 192–193 USDOT/FHWA; 194, 195 © Bettmann/ CORBIS; 196 USDOT/FHWA; 197 © Bettmann/CORBIS; 198 © Richard T. Nowitz/CORBIS; 199 courtesy Parsons Brinckerhoff; 200 left © David Sailors; 200 right courtesy Maryland Transportation Authority; 201 top © David Sailors; 201 bottom courtesy Maryland Transportation Authority; 202 © Dan McNichol; 203, 204, 205, 206, 207 © Central Artery/ Tunnel Project; 208 © Dan McNichol; 209 © Dan McNichol (top), Central Artery/Tunnel Project (bottom); 210 © Dan McNichol; 211 © Dan McNichol; 212 USDOT/FHWA; 213 Central Artery/Tunnel Project; 214 © David Sailors; 215 Central Artery/Tunnel Project; 216 © David Sailors; 217 Charlie Archambault for *US News and World Report;* 218–19 © David Sailors; 221 © Will & Deni McIntyre/CORBIS; 223 © Sandy Felsenthal/CORBIS; 224 © David Sailors; 227 top © Mark E. Gibson/CORBIS; 227 bottom © John MacPherson/CORBIS; 229 © Dan McNichol; 230 © Martyn Goddard/CORBIS; 232 USDOT/FHWA; 233 left © David Sailors; 233 right USDOT/FHWA; 234 © David Sailors; 236 courtesy Georgia Department of Transportation; 238 courtesy WesTrack; 239 courtesy PaveTrack/Jim Killian; 240, 241, 242, 243, 244 © Central Artery/Tunnel Project/Phil DeJoseph.